citizen environmentalists

CIVIL SOCIETY: HISTORICAL AND CONTEMPORARY PERSPECTIVES

Series Editors:

Virginia Hodgkinson
Public Policy Institute
Georgetown University

Kent E. Portney
Department of Political Science
Tufts University

John C. Schneider
Department of History
Tufts University

For a complete list of books that are available in the series, visit www.upne.com.

James Longhurst, *Citizen Environmentalists*

Bruce R. Sievers, *Civil Society, Philanthropy, and the Fate of the Commons*

Janelle A. Kerlin, ed., *Social Enterprise: A Global Comparison*

Carl Milofsky, *Smallville: Institutionalizing Community in Twenty-First-Century America*

Dan Pallotta, *Uncharitable: How Restraints on Nonprofits Undermine Their Potential*

Susan A. Ostrander and Kent E. Portney, eds., *Acting Civically: From Urban Neighborhoods to Higher Education*

Peter Levine, *The Future of Democracy: Developing the Next Generation of American Citizens*

Jason A. Scorza, *Strong Liberalism: Habits of Mind for Democratic Citizenship*

Elayne Clift, ed., *Women, Philanthropy, and Social Change: Visions for a Just Society*

Brian O'Connell, *Fifty Years in Public Causes: Stories from a Road Less Traveled*

Pablo Eisenberg, *Challenges for Nonprofits and Philanthropy: The Courage to Change*

Thomas A. Lyson, *Civic Agriculture: Reconnecting Farm, Food, and Community*

Virginia A. Hodgkinson and Michael W. Foley, eds., *The Civil Society Reader*

Henry Milner, *Civic Literacy: How Informed Citizens Make Democracy Work*

Ken Thomson, *From Neighborhood to Nation: The Democratic Foundations of Civil Society*

Bob Edwards, Michael W. Foley, and Mario Diani, eds., *Beyond Tocqueville: Civil Society and the Social Capital Debate in Comparative Perspective*

Phillip H. Round, *By Nature and by Custom Cursed: Transatlantic Civil Discourse and New England Cultural Production, 1620–1660*

Brian O'Connell, *Civil Society: The Underpinnings of American Democracy*

citizen environmentalists

JAMES LONGHURST

TUFTS UNIVERSITY PRESS
MEDFORD, MASSACHUSETTS

Published by
UNIVERSITY PRESS OF NEW ENGLAND
HANOVER AND LONDON

Tufts University Press
Published by University Press of New England,
One Court Street, Lebanon, NH 03766
www.upne.com

Printed in U.S.A.
5 4 3 2 1

Library of Congress Cataloging-in-Publication Data
Longhurst, James Lewis.
Citizen environmentalists / James Longhurst.
p. cm. — (Civil society : historical and contemporary perspectives)
Includes bibliographical references and index.
ISBN 978-1-58465-849-8 (cloth : alk. paper) — ISBN 978-1-58465-859-7 (pbk. : alk. paper)
1. Environmentalism—Pennsylvania—Pittsburgh—History—20th century. 2. Environmentalism—United States—Pittsburgh—History—20th century. 3. Air—Pollution—Pennsylvania—Pittsburg—History. 4. Air—Pollution—Pennsylvania—Pittsburg—History. 5. Environmental policy—Pennsylvania—Pittsburgh—Citizen participation—History. 6. Environmental policy—United States—Citizen participation—History. 7. GASP (Pittsburgh, Pa.)—History. I. Title.
GE198.P4L66 2010
363.7'05250974886—dc22 2009045856

University Press of New England is a member of the Green Press Initiative.
The paper used in this book meets their minimum requirement for recycled paper.

FOR JENNIFER, WHO SHOWED ME
THE VIEW FROM MOUNT WASHINGTON

Contents

Illustrations follow page 111.

Preface

THE CITIZEN ENVIRONMENTALISTS

A funny thing happened on the way to study the public forum: I didn't find what I expected. As a historian interested in public debate, I've long been working under the loose thesis that when people speak in the public sphere on matters of policy, they give away more than they mean to say. The language they use, the arguments they make, the examples they choose—all of these things betray, intentionally or not, the mindset of the speaker. We can see their assumptions, their intentions, and above all the ways that they conceive of the society around them. When arguing about the most mundane policy matters—roads, infrastructure, sewers, schools—participants in public debate give away their vision of the proper functioning of society; their understandings of the meaning of race, class, and gender; the goals of government or the meanings of capitalism. When more disruptive topics deeply divide the public, historians can see even more. Like a bulldozer that has carved a deep trench in the earth, disruptive public debates expose the layered sediment of assumptions hidden from view in more quiet times. I had previously studied the topic of divorce law and reform attempts using this operating thesis. The language and assumptions of speakers on the meaning of marriage was most illuminating.

I was thinking about public debate in this way in the late 1990s when I moved to Pittsburgh, Pennsylvania, to begin this project. In Pittsburgh the biggest public debate concerned the city's air pollution and attempts to forge a viable future amidst the crumbling remains of an industrial past. This debate had flared up once again at just the moment that I began searching for a research topic. When I went to a public meeting called to discuss a proposed revival of an industrial site that promised both more pollution and more jobs, I was treated to the sight of these debates in the flesh. On one side were diminutive, middle-aged professional women, housewives, and mothers who represented various civic and neighborhood groups, defending their vision of a city as a place to live and to raise a family. On the other side were the men of the local labor unions, in their union jackets and work boots, defending the simple observation that living and raising a family required a job and a paycheck. The two sides presented their cases in front of the microphone and then argued with each other, squaring off, toe-to-toe. The working-class men loomed over the clubwomen, who nevertheless did not give an inch. This, I thought, was the jackpot: a real live local debate that would expose all sorts of interesting assumptions about jobs, the environment, and the proper role of government.

But when I began writing a history of environmental debate in postwar Pittsburgh, I was expecting to find the local version of what historians said was happening on the national level. I expected to find an upswing in popular interest in environmental philosophy, care for the earth, and ecological concerns surrounding the first Earth Day in 1970, possibly concentrated among college students and mixed-gender groups inspired by campus protest and the New Left. Instead, I found that activists were using language and organizational tools that had more to do with Progressive Era organizing, maternal care for the environment, expert knowledge, and a recurring emphasis on the rights and responsibilities of citizenship. Thinking that this was simply an aberration caused by Pittsburgh's oft-noted social and cultural conservatism, I was again surprised to find that these themes appeared across the nation. The thing is, these new environmentalists spent as much time talking about citizenship and maternal care for children and society as they did discussing the logic of environmental philosophy; they identified themselves as *citizens* and *mothers* long before they took the title of *environmentalists*; they spent a great deal of time dealing with the problems of neighborhood organizing, participatory democracy, and the rights of citizens to intervene in policy matters. At least on the local level, members of the new, modern environmental movement weren't as environmental as one might expect.

From this perspective, it would be more correct to say that there was a nationwide phenomenon best named citizen environmentalism, that these activists benefited from similar rhetoric and social capital, and that they are best understood by examining their varying degrees of substantive impact, made possible both by new federal requirements and by their own actions. These new local activists were the unnamed, widespread, and diffuse environmental groundswell. They were not necessarily the campus-based student movement associated with the New Left, though they were inspired by it. They were not the same thing as the nationally organized conservationist groups, though they occasionally interacted with them. They were not identical to the civic organizations of the Progressive Era, though they were occasionally similar. Most importantly, they were not the same thing as the national environmental organizations, which due to their ubiquity and visibility have dominated the story of the environmental movement. Recovering the stories of these small and scattered citizen groups should cause us to reexamine our narrative of the modern environmental movement. In this book, I begin the first task and offer some observations on the second.

• • •

At the tail end of the social upheaval of the 1960s, something unprecedented happened in the United States: large numbers of Americans began to express a new interest in environmental matters. We call this upwelling the modern environmental movement, and many accounts of this event focus on the precedents and emerging philosophy of this new trend. But these aspects of the story do more to show the continuity of events with the past than they do to explain the phenomenon of a broad surge in environmental interest. In other words, in telling the history of the modern environmental movement, environmental historians have focused on the *environment* to the exclusion of the *movement*. What was new about the movement, and what makes it historically significant, was not that Americans were discussing environmental concerns. They had been doing so in resource, land use, conservation, industrial hygiene, pollution, and plant and animal management disputes for centuries. Rather, what was new was that the legacy of recent social movements equipped environmentalists with the legal, political, and rhetorical tools of individual protest, participatory democracy, and interest-group lobbying. The environmental concerns themselves were not new; an active definition of citizenship was.

What would happen if we were to consider the environmental movement as a social movement—to question who, exactly, made up the membership and leadership of the new environmental organizations, how they worked to accomplish their goals on a day-to-day basis, and what had changed to make this broad-based movement possible? If we were to take a social historian's approach to these questions, the result would be an increased focus on the demographics and mechanisms of local participation in the movement rather than on national politics or philosophical trends. Three observations become visible with this approach: First, that there was a widespread trend of a specific type of local organizing in the early 1970s, one that emphasized citizenship and participatory democracy and that created a newly diffuse layer of similar but independent groups in cities, neighborhoods, and regions across the United States. Second, that these organizations depended upon the rhetoric of citizenship, maternal care for the environment, and professional and scientific expertise while benefiting from the social capital of their largely white, middle-class, educated memberships. And third, that analysis of these groups should focus on determining their substantive impact on both environmental policy and the inclusion of citizens within policy implementation and enforcement, proceeding in a cyclical, dialectical process between local pressure and federal action.[1]

These three themes can be used to examine the boom in public involvement in environmental matters through events in cities across the nation. Observers often refer to this increase in public discourse and action, centered around the first Earth

Day in 1970, as the modern environmental movement. But historians have failed to probe below the national-level events, large organizations, and federal policies in explaining the causes and consequences of this movement. In fact, the boom in public involvement occurred on the local level, fueled by small, fleeting, ad hoc organizations. While these neighborhood and city groups may have consisted of no more than a telephone number, a mimeograph, and a few motivated individuals, their existence demonstrates a fundamental transformation in the importance and efficacy of a particular type of citizen involvement. These groups were the individual components of the groundswell that was the environmental movement of the late 1960s and early 1970s. Their foundations in citizen rights, women's organizational work, and government policy deserve a closer look.

• • •

For many observers, the ability of small groups of otherwise nonpolitical Americans to organize for political goals has a symbolic importance that far outstrips the actual numbers and accomplishments of the activists themselves. As such, when some social scientists discuss neighborhood organizations meeting in the basement of a local church or in the community room of an elementary school, they may actually have far weightier concerns in mind: the fate of the republic, or of nations. Thus, any analysis of environmental activism connects with a variety of debates in political science, environmental history, and the history of postwar America. Specifically, the analysis of the rise of the citizen environmentalists presented here uses the language and theoretical models of political scientists who study the civic engagement debate to offer a counternarrative to the pessimistic views of those who consider the last quarter of the twentieth century a wasteland of lobbying, populated by interest groups and "associations without members." In addition, this approach complicates the narrative presented by environmental historians of the rise of the modern environmental movement, bypassing the traditional questions of *why* the movement arose for questions of *how* the movement gained power. Finally, this emphasis on the importance of citizenship and rights for early environmentalists should cause us to reexamine the disciplinary boundaries historians have placed between those who study the environmental movement and those concerned with all of the other social movements of the nearly two decades of protest and upheaval known as the long 1960s.

The role of the individual in relation to government has always been of overwhelming importance for political scientists, historians, social scientists, reformers, politicians, and general busybodies. This is an unsurprising development in a nation that prizes the rights of individuals and promotes the importance of democratic values. Historians have focused on civic organizing during and after the revolutionary period, throughout the early Republic and again in the first decades of the twentieth century.[2] Scholars study local expressions of various national movements, including revolutionary politics, the abolitionist crusade, suffrage movements, and the various social reforms of the Progressive Era. Civic reformers from John Dewey to Bill Clinton have variously diagnosed and attempted treatment for perceived civic decline.[3] For their part, political scientists have flocked to the concerns of nineteenth-century observer Alexis de Tocqueville as a starting point for extended debates about the efficacy, methods, theory, and impact of civic associations with political goals. Variously referred to as the "Tocqueville" or the "Civil Society" or the "Civic Engagement Debate," these discussions use historical evidence, theoretical models, and contemporary observations to come to differing conclusions about the power, success, and future of individual involvement in public affairs.[4]

The central theme that runs through all of these works on citizenship is a concern for the proper functioning of a democratic society. For nearly all of these observers, whether and how individuals organize themselves to influence government is a critical indicator of the success of the nation on a grand scale. Writing in 1967, political scientists Edward C. Banfield and James Q. Wilson noted that "Citizen participation . . . has in recent years come to be regarded in many quarters as a normative principle inseparable from the idea of democracy itself." There are two interesting implications here. First, democracy and citizen participation came to mean the same thing in the minds of Americans. Second, this concept changed over time. According to some, citizen participation, public involvement, and participatory democracy became more important in the late twentieth century.[5]

Three concepts that have dominated this discussion—the concept of the "public sphere," measurements of "social capital," the geopolitical importance of "civil society"—demonstrate this linkage between individual action and the health or viability of the entire society. Many theoretical approaches to these interactions of citizens have come to use the concept of a "public sphere," or a conceptual space for public interaction and expression outside the formal mechanisms of government. According to theorists, less authoritarian or paternalistic societies allow standardized debate or activity in public space, possibly facilitated by legal, cultural, or journalistic institutions. While not as destructive as a riot, such public space offers a similar opportunity to circumvent a clogged, unresponsive, or unrepresentative bureaucracy. Theorist and sociologist Jurgen Habermas writes that "The bourgeois public sphere may be conceived above all as the sphere of private people come together as a public" and that this newly assembled people may use the public sphere "against the public authorities themselves, to engage them in a debate over the general rules governing relations" in society. Despite the fact that Habermas was referring only to seventeenth and eighteenth century Europe in this description of the public sphere, his definition has been widely applied by theorists and academics across multiple disciplines to describe nearly all public action.[6]

The work of sociologist James Coleman has likewise been appropriated to describe the importance of individual political organizing. Coleman theorized that a person's "social capital"—that is, access to social ties, networks, and shared norms—could determine that individual's capacity to function within a society. Following Coleman, political scientist Robert Putnam used the concept of social capital in a series of prominent works. In brief, Putnam argued that high levels of social capital translated into highly effective participatory democracies and vital economies. The more socially interconnected the populace through voluntary organizations, churches, social groups, or networks, the more efficient and effective their government. Likewise, the less social capital an individual enjoys, the less ability that individual might have to receive equal political representation, fair treatment in a justice system, or access to economic opportunity. Given that basis, Putnam's groundbreaking argument in the mid-1990s that the interactions that had previously created social capital were on a decades-long decline in the United States drew a great deal of popular and scholarly attention. If Americans were now "bowling alone" rather than with their neighbors, would it mean that the nation would function less efficiently, or cease to function entirely?[7]

These measurements of social capital have garnered attention because of the perceived importance of civil society. If the public sphere is one label for the location of nongovernmental power, then "civil society" is a much more expansive definition for decision making, innovation, organization, and action outside of the direct control of governments or corporations. The concept of civil society includes all voluntary organizations and institutions outside of government control and the vital and synergistic interactions between such entities. To visualize the impact of civil society, imagine a campaign—let's say to eliminate smoking, or to curtail political support for a repressive foreign regime, or to promote a specific domestic policy. This campaign might include interactions between several public interest groups, some socially active churches, several newspapers, a longstanding charitable institution like the American Red Cross, a loosely organized group of political bloggers, and a nonprofit research institution such as a medical school or university. Together, these institutions might form a significant force for change, but none of them would represent the power of the government or of a single corporate interest. They would instead be defined as a component of civil society. For many observers, civil society has come to mean a civic culture that allows, supports, and sustains such vital interaction outside the formal institutions of democratic government, on the level of both nations and cities. Many have returned to Tocqueville's musings about the purpose and effect of civil association, describing them as powerful counterweights to state power.[8] Others have begun to posit civil society as a potent force in the democratization or westernization of nations, a deeply held concern in the era of the cold war and in subsequent years. These types of questions have spurred a great deal of speculation about the cause, makeup, demography, and effects of civil associations, in some cases focusing specifically on community-level environmental activism.[9]

Political scientists have dominated the study of civic engagement and civil society, even when the object of their study has been historical. This is partly so because the topic has been a natural concern for political scientists. But it's partly for another reason: while historians have examined earlier movements, political scientists have spent comparatively greater effort in examining the more recent events and movements of the 1960s and 1970s. As archival sources become available, it is quite likely that historical work in this area will catch up to the theoretical and analytical work of political scientists in this area. But for the moment, political scientists dominate the study of civic and neighborhood organizing. This creates the ironic development that, while most historians have not been drawn into the arguments concerning civil society, their data and conclusions have. To test the hypothesis of a decades-long decline in social capital, political scientists have branched out from their traditional study of voter rolls to inquire into civic organizing, church attendance, kinship networks, and community studies—in other words, the interests of social historians. In particular, histories of women's organizations, national reform movements, and Progressive Era civic and social clubs have been used to demonstrate relative levels of social capital. Historians have demonstrated both change and continuity in examining civic activism, arguing that there is ample evidence of organizing and action in many historical eras and that there is a change in the type of engagement throughout the twentieth century as citizens found ways of insinuating themselves into power relationships that had previously been reserved for politicians, professionals, and industrialists. Political scientists claim that citizen presence in policy debates remade the method and practice of policymaking in the United States after World War II and may represent a new direction in democratic organizing and the interactions of state and society.[10]

The civic engagement debate makes small-scale political organizing into a significant measurement of a just and democratic society. If we follow that logic, the modern environmental movement serves as a case study in civic engagement, a test of whether and how citizens without immediate access to political power nonetheless could organize to influence decisions and events. For those involved in these debates, the relative success or failure of small organizations of housewives, retirees, and part-time enthusiasts in lobbying for environmental reform takes on a whole new significance.

If the civic engagement debate furnishes political significance for the study of the citizen environmentalists, what then is the historical significance of the broad surge of environmental activism in the United States during the late 1960s and early 1970s? This question, which one might think central to the field of environmental history, has received insufficient attention and lacks a satisfying answer to guide this book's analysis of the citizen environmentalists. Save for that of a few prominent writers such as Samuel Hays, Robert Gottlieb, and Adam Rome, there has not been extensive work on this question. Indeed, an outsider to the field of environmental history might be surprised to hear that only a bare minimum of historians pay any attention to the U.S. environmental movement of the long 1960s. But it's true; very few environmental historians write about the environmental movement. There are other topics that pique the interest of environmental historians, including the complexities of scientific debates, the causes of technological change, shifting streams of philosophy and cultural expressions of relations with nature, the development of government policy, gendered discourse, and economies of commodities and resources.[11] These approaches are often chronologically distant from the movement of the 1970s, as historians often explore these themes through events and sources centuries removed from the first Earth Day. Finally, those historians who do examine the environmental movement often do so from the national level.[12]

This book argues that the significance of the modern environmental movement lies in the changing definition of citizenship in relation to an increasingly powerful state. The "green revolution" on the national level was built upon the "rights revolution" that preceded and accompanied it, equipping small, local citizen's groups with knowledge of legal rights, media tactics, organizational models, and rhetorical approaches. These small groups—the so-called grass roots—have not yet been the subject of sufficient historical research.[13] Closer examination of these groups uncovers other themes, specifically concerning their organization, their use of language, and their relative levels of success. This analysis puts the environmental movement in context with other social reform movements of the 1960s and demonstrates that the surge in such groups occurred not simply because of a spontaneous increase in American's interest in environmental concerns but also because of the enabling features of a decade of citizen activism. In other words, the significance of the environmental movement lies in the movement itself.[14]

Taking this argument as given—that the environmental movement of the 1970s should be understood as a single episode in a wide-ranging revolution in society and politics—what does this imply for the study of the environmental activists of that period? By adding the context of the 1960s we expose a number of new approaches to understanding the modern environmental movement, including a focus on local action, an increased understanding of the existing institutions and social capital that made the movement possible, and an examination of the organizational tools and legal rights available to organizers and activists. Viewed from this perspective, the community-based environmental activism of the long 1960s was only one of many attempts to redefine an individual's place in relation to the state. As such, local environmental activism owed much to similarly organized groups in the civil rights, women's liberation, antiwar, ethnic identity, and community organizing movements. All of these movements became publicly prominent in the decades of the 1960s and 1970s. All eschewed the traditional structure of party politics in exchange for "identity politics." All used demonstrations, organizing, and media coverage to encourage public involvement in their movement of choice. But most importantly, all advocated increased involvement by individuals in their government and society.[15] The study of environmental organizing could thus benefit from an understanding of the many other movements of the period, the historiography of the 1960s and an increasing interest in the decade of the 1970s.[16]

Research into civil rights organizing has advanced far beyond what has been accomplished in the history of environmental organizing, and reference to this work points environmental historians toward the importance of local organizing. To use the language of the civic engagement debate, the driving force of the civil rights movement was the activists' use of social capital within black communities to mobilize existing resources in the service of the movement. Aldon D. Morris has argued that aspects of resource mobilization theory can describe the origins of the civil rights movement. "The basic resources enabling a dominated group to engage in sustained protest," says Morris, "are well developed internal social institutions and organizations . . . that can be mobilized to attain collective goals."[17] Morris's focus on local action inspired a large number of community studies, as well as an emphasis on the organizational abilities and institutions that existed in black communities before the height of the movement. For example, Belinda Robnett particularly highlights the importance of women's organizational work in mobilizing communities.[18] This model of mobilization proved hugely successful and Morris argues that it inspired subsequent generations of organizers—including, one would assume, environmentalists—to emphasize organization outside the boundaries of traditional electoral politics: "In a loud and clear voice the civil rights movement demonstrated to those groups that organized nontraditional politics was a viable method of social change, capable of bringing about the desired results far faster than traditional methods." Indeed, writes Morris, "the modern women's movement, student movement, farm worker's movement and others of the period were triggered by the unprecedented scale of nontraditional politics in the civil rights movement." This thesis has become commonplace, as historian of citizenship Michael Schudson has noted that "the civil rights movement provided a model and inspiration for a wide array of new social movements and political organizations." If this view is correct and all of the subsequent movements of the rights revolution were triggered by the example of the civil rights movement, then shouldn't we attempt to understand environmental activism in the same way that we understand the civil rights organizations?[19]

There are, of course, as many comparisons as there are social movements in the 1960s. In the same manner that civil rights organizations used social capital to capture national headlines with local protest in the expanding civil rights movement, the women's movement relied upon local organizations outside traditional party politics to exert pressure for social and political change. Particularly in the case of the state-by-state ratification of the Equal Rights Amendment in the 1970s, the women's movement found it necessary to use the same sort of social capital to bolster local support for a national project. The national campaign benefited from demonstration, protest, and issue-driven community organizations on the local level to redefine the ways in which women were politically represented on the national level. The women's movement was no monolithic organization, but a loose agglomeration of tens of thousands of locally organized, action-oriented groups: self-help clinics, battered-women's shelters, rape-crisis centers, consciousness-raising groups, women's studies programs, and political action lobbies.[20] As Estelle B. Freedman notes, "breaking away from the male-dominated New Left, the women's liberation movement created a network of predominantly white, women-only organizations." Those multitudes of local, autonomous organizations followed their own strategies toward their own goals, but together made up a perceptible movement on the national level. As such, these many constituent organizations and struggles that made up the greater movement provide yet another point of comparison and approach to understanding the environmental movement.[21]

Adam Rome has in fact recommended such an approach in his article "'Give Earth a Chance.'" Rome calls on historians to look to the social, cultural, and political trends of the period—"the revitalization of liberalism, the growing discontent of middle-class women and the explosion of student radicalism and countercultural protest"—for an explanation of the "explosive growth" in environmental concern.[22] In effect, Rome's essay might be an application of the trends outlined by Morris and others: close analysis of the demographics, concerns, and evolution of small-scale components of the environmental movement in order to more fully understand the chronology, causation, and historical context of the larger, national movement.

This trend of examining the many social movements of the long 1960s as interrelated, syncretic, and synergistic can be labeled the "rights revolution" approach. The benefits of this trend are legion. It links the shared characteristics and contexts of the many movements. It provides important grounding for our understanding of each, making a coherent whole out of an otherwise fractured and fractious decade. And there is particular utility here for understanding the environmental movement as a product of its time.[23]

There are intellectual drawbacks, however: contextualization of social movements in foreign policy, national legislation, and cultural zeitgeist might serve to diminish the importance of individual leaders, the phenomenon of mass protest, or the sacrifices of heroic activists. And it makes strange bedfellows out of disparate groups, linking civil rights and the Stonewall riots, women's liberation and Barry Goldwater, Cesar Chavez and Boston's antibusing activism. This study should not be read as conflating these movements. Neither does it seek to diminish the individual suffering, moral struggle, unspeakable violence, and national upheaval that were components of the freedom struggle. In comparison, the environmental movement was relatively bloodless, a polite legal and political disagreement largely experienced in courtrooms and legislatures rather than the streets. The impact and goals of these movements are obviously not comparable, but the tools and strategies available for organizing certainly are.

While environmental activists were not explicitly modeling their strategy on the moral arguments of the civil rights movement or on any of the other movements of the period, the opportunities for their active political involvement were nonetheless made possible by the strategies of previous movements that highlighted the citizen's equal right to substantive inclusion in political processes. Together, the examples of local organizing from the civil rights and women's movements—and countless other community organizations born in the 1960s—gave impetus to a similar surge of neighborhood or community organizing for environmental goals. The resulting model for pluralistic governing drew from and influenced similar organizing and involvement nationwide; but it was a part of a broad-based transformation that was not limited to environmental politics. It was instead merely a component of the rights revolution.

• • •

This book uses a variety of primary documents from local, state, and national sources to demonstrate the circumstances, rhetoric, strategy, and effects of increased citizen involvement in environmental policymaking, implementation, and enforcement. These documents include legal records, journalistic accounts, and previously unavailable internal documents from citizen's groups. Using these records, Chapter 1 argues that the local environmental movement of the early 1970s reflected nationwide trends encouraging public involvement and redefining citizenship to include an active public. Chapter 2 turns to the history of Pittsburgh's lack of public involvement in policymaking until the 1960s and the example of Pittsburgh's Group Against Smog and Pollution to demonstrate this transition to an inclusive and confrontational model on the local level. This chapter uses the legal concept of "standing" to examine how GASP fought to gain a *legal* right to represent the public interest on matters of air pollution control and a *rhetorical* right to represent that interest in the court of public opinion. Chapter 3 examines the rhetoric, demographic makeup, and social capital of the new wave of environmental groups in Pittsburgh and across the nation, and argues that GASP was an exemplary case of many such groups becoming active at the same time. Chapter 4 demonstrates the effectiveness of the gendered rhetoric and maternalist urban reform vision embodied in GASP's educational mission. Chapter 5 demonstrates the very real, but ultimately constrained, impact of the reformed policy implementation and enforcement process. This chapter focuses on the county's Air Pollution Appeals and Variance Review Board, which became a nexus of public debate for representatives of industry, government, and the citizenry. Pittsburgh's remarkable pool of independent technical and scientific experts raised the level of debate using the venue of the Variance Board. Once the enforcement attempts of the board reached their limits, citizen's groups attempted to push legal action against two very large industrial corporations, United States Steel and Jones & Laughlin. Chapter 6 narrates this long-running battle, arguing that public involvement helped instigate and maintain this messy, convoluted, decades-long saga of enforcement. Finally, the conclusion returns to the significance of the citizen environmentalists for the modern environmental movement, Pittsburgh, and GASP itself.

• • •

While any error in this manuscript is certainly my own, any success I have had with this project is the result of the support, advice, and guidance I have received from countless sources. I thank them here individually and collectively; if I have overlooked anyone, I apologize.

Many of my colleagues have read this work at various stages, beginning as a dissertation. First among these is Joel Tarr. I chose to work with Joel not only for his intelligence and accomplishments but also for his kindness and humanity. I treasure the time spent in his office and thank him for suggesting, in the spring of 1998, that I attend a public meeting sponsored by some group with a funny acronym for a name. John Soluri gave an incredible amount of his time to repeatedly read every sentence and footnote of the dissertation. I wish to particularly thank Maureen Flanagan, whom I shanghaied mercilessly. One minute she was minding her own business at an Urban History Association meeting, the next she was serving—with graciousness and insight—on my committee. Several other colleagues have read part or all of this work and provided immensely helpful commentary, including Elizabeth Blum, Bob Burke, Steve Corey, Jennifer Trost, and Greg Wilson. Carl Zimring has read countless versions of this project since 1998 and for that, I apologize. Vagel Keller and I collectively make up what is probably the smallest, but also the best, graduate cohort in the world. I appreciate the friendship and support of my colleagues in the history departments at Muskingum College and the University of Wisconsin-La Crosse; without their moral and logistical support, this book would not have made it this far.

Historical writing stands or falls on its foundation of primary documents, which would not exist without the ceaseless efforts of innumerable archivists. My work is no exception. I appreciate the assistance of Dominic LaCava, Kate Colligan, Alesha Shumar, Miriam Meislik, and Mike Dabrishus at the Archives Service Center of the University of Pittsburgh; Audrey Iacone and, later, Gilbert Pietrzak at the Carnegie Library of Pittsburgh; Joe Schwarz at the National Archives; Roger C. Westman and

Catherine Garner at the Allegheny County Health Department; Jonathan R. Stayer at the Pennsylvania State Archives; Angelica Kane at the *Pittsburgh Post-Gazette*; and Joel Fishman at the Allegheny County Law Library.

I've bounced ideas in this book off of the active and inventive minds of undergraduate students in my environmental history courses at Carnegie Mellon, Muskingum, and UW-La Crosse. One of the joys of my teaching career is to meet the students from across the campus who go out of their way to take a class with *environment* in the title; their natural excitement and commitment reinvigorates me when it all seems so gloomy, and Matt Harris gets a shout-out here for his research contribution.

The editorial and production staff at the University Press of New England, including Ellen Wicklum, Phyllis Deutsch, Amanda Dupuis, Lori Miller, and Lys Weiss, has been amazingly helpful in the preparation of this manuscript. Thanks especially to eagle-eyed copyeditor Robert Milks.

This project would not have been possible without the many dedicated volunteers of GASP, past and present: Kate St. John, Rachel Filippini, and Sue Seppi today carry on a forty-year long tradition of brilliant and tireless women at GASP. Walter Goldburg is the only founding member of the group to still be an active part of the board of directors. Thanks to Marie Kocochis, Fran Harkins, and Jonathon Nadle. Michelle Madoff is still "peppery," opinionated, outspoken, and politically active; I'm honored she would speak with me and welcome me into her Arizona home. Jeannette Widom was happy to sit down and swap stories of GASP; she was the first of several GASP activists to do so and I thank her for it. Bernie Bloom has been an invaluable resource in understanding the contingencies and people of Pittsburgh and of environmental activism; I appreciate our spirited discussions and respect his principled disagreements with my observations. Emeritus Professor David Fowler of Carnegie Mellon, a former GASP board member himself, read an early version. Nine years later, new GASP legal director Joe Osborne read chapters descended from that prospectus. I thank them all. For purposes of clarity and openness, I should declare here that I am a former member of the board of directors of GASP and a current member of its board of advisors. GASP assisted this project by allowing access to its internal records. However, GASP has not had authorial input or editorial control over this project; all opinions and conclusions herein are my own.

Acronyms and Abbreviations

ACAPCAC	Allegheny County Air Pollution Control Advisory Committee
ACCAAP	Allegheny County Citizens Against Air Pollution
ACCD	Allegheny Conference on Community Development
ACHD	Allegheny County Health Department
AQA	Air Quality Act
BAPC	Bureau of Air Pollution Control (of the Allegheny County Health Department)
CAA	Clean Air Act
CAC	Citizen's Advisory Committee
CAP	Community Action Program
CETF	"Citizens" Environmental Task Force
CHOC	Citizens Helping Our Community
COKE	Clairton Organizes to Keep Employment
DER	Department of Environmental Resources
EIS	Environmental Impact Statement
EOA	Economic Opportunity Act
EPA	Environmental Protection Agency
EQB	Environmental Quality Board
ERAP	Economic Research and Action Project
GASP	Group Against Smog and Pollution; also several other groups outside Pittsburgh, with various arrangements of words to create the same acronym
GFWC	General Federation of Women's Clubs
GRIP	Group Recycling in Pittsburgh
HEW	Health, Education and Welfare
IRS	Internal Revenue Service

J&L	Jones & Laughlin Steel
LTV	Ling-Temco-Vought Corporation
NAM	National Association of Manufacturers
NAPCA	National Air Pollution Control Administration
NARA	National Archives and Records Administration
NCCS	National Center for Charitable Statistics
NEPA	National Environmental Policy Act
OAP	Office of Air Programs
OPA	Office of Public Affairs
OSHA	Occupational Safety and Health Administration
PHS	Public Health Service
PIRG	public interest research group
PPG	parent company of Pittsburgh Paints and Pittsburgh Plate Glass
PSAPCA	Puget Sound Air Pollution Control Agency
SCRAP	Students Challenging Regulatory Agency Procedures
SING	Survival with Industry Now through Growth
SIP	State Implementation Plan
USC	United Smoke Council
USWA	United Steelworkers of America
VISTA	Volunteers in Service to America
WHPA	Women's Health Protective Association
WSP	Women Strike for Peace

citizen environmentalists

1 Power to the Public Hearing

THE IMPORTANCE OF CITIZENSHIP TO THE ENVIRONMENTAL MOVEMENT

"What's the difference between a professional citizen and a citizen?"

—Maurice K. Goddard, Secretary of Pennsylvania Department of Environmental Resources

"A professional citizen is one who works at it full time and studies how to be one."

—Joan Hays, self-declared professional citizen, January 17, 1972

It was supposed to be a quiet meeting, but it turned into the political equivalent of an ambush. The topic of the day wasn't anything new, and there was no pressing deadline; after all, it was just another advisory board meeting to discuss a proposed plan to clean the perpetually dirty air in Pittsburgh, Pennsylvania. For more than a hundred years Pittsburgh had been known as the "Smoky City," and for the latter half of that century the city's residents had engaged in repeated attempts to clean the air, with varying degrees of seriousness and success. During that time, most decisions about controlling smoke and air pollution had been made behind closed doors. Compromises had been negotiated this way—in private, between local government and industrial leaders—for decades, through half-hearted attempts to voluntarily limit smoke in the first quarter of the century, throughout a visibly successful campaign to control smoke immediately after World War II, and in a number of policy revisions at the state and county level since.

But the meeting of September 24, 1969, turned out to be very different. The Allegheny County Board of Health had originally planned to discuss changing local air pollution control laws for, perhaps, a few hours. Instead, the meeting stretched over three days, and was filled with acrimony, tears, and public denunciations. An event that might normally attract less than a dozen public comments swelled to include hundreds of citizens, civic organizers, public health officials, housewives, steelworkers, and journalists from both print and electronic media. County officials scrambled to find an auditorium large enough for the five hundred Pittsburghers who wanted to sit in the audience on the first day. That first session was delayed "when about 150 latecomers could not gain admittance to the already packed hearing and set up an angry clamor in the hall outside." The several hundred who made it into the room arrived with prepared statements to inform the committee members that their proposed air pollution plan was not only poorly thought-out and impractical but also represented the "legalized murder" of steelworkers. For three days a "steady stream of angry Allegheny County residents" rose to speak at a single microphone, facing the eleven members of the committee from the meeting hall floor. Former steelworkers stood up to read typewritten paragraphs describing their emphysema. A mother presented her four-and-a-half month old daughter and said that because of air pollution, the wailing baby's "lungs were sore and her eyes ran red." Citizens appeared wearing buttons emblazoned with the skull and crossbones, the words "Cough, cough!" or "Remember Donora!" in reference to a deadly air pollution disaster. Law students wore surgical masks. The county's chief forensic pathologist reported his observations linking air quality and increased mortality rates. A county commissioner called for the "banishment" of industrial representatives from all future hearings "in all but a second class status," and later accused automakers of "criminal negligence." Speakers called for jail terms for the "boards of directors of guilty polluting firms." "Citizens Turn Air Blue" said one headline, "Citizens Flay Anti-Pollution Plans" another. Local papers summarized the meetings by reporting that "the roof fell in" on the committee members, that the proposed criteria "took a beating," and that the public had "ripped them to shreds."[1]

This was not a spontaneous expression of anger. The citizens of Pittsburgh had spent months preparing for this one meeting, in a loosely organized attempt to transform not only local air pollution control rules but also the methods by which those rules were created. At planning meetings for months in advance, area residents were coached by leaders of the League of Women Voters, the Federation of American Scientists, a local group known as the Western Pennsylvania Conservancy, and a dissident member of the advisory committee. These organizations helped prepare printed testimony, line up a wide variety of speakers, and foster the creation of a core group of activists—a coalition of civically active housewives, local academics and scientists, and other interested parties. They seemed to pop up overnight; as a local newspaper put it in 1970, "citizen's groups against pollution have blossomed like flowers in unpolluted soil." The *Wall Street Journal* described this core group as "The Breather's Lobby"; *Business Week* called them "citizen's groups and conservationists"; the local press occasionally called them "Clean Air Crusaders." One assemblage of Pittsburghers eventually chose to call themselves the "Group Against Smog and Pollution," or GASP. But wherever they were in the nation, these organizations might be described for the purposes of this book as "citizen environmentalists."[2]

The surprisingly contentious public hearing described above was an example of a new type of activist environmentalism in the United States, one that emphasized the involvement of citizens in government. Throughout the decade of the 1970s, the new movement was confrontational rather than accommodating, promoted public debate rather than private agreements, was intent upon litigating mandatory change rather than voluntary improvements, and focused on substantive public involvement. A variety of forces worked to emphasize both the symbolic and the substantive importance of public hearings and public involvement in policymaking. The result was a new type of decision-making system, one that was influenced by the popular language of participatory democracy, dependent upon newly evolved legal definitions of the rights of citizens, and spurred by federal requirements for public involvement in state and local decision making. In the fall of 1969, the people of Pittsburgh were about to become citizen environmentalists, seeking substantive involvement in political matters by using whatever rhetorical justifications and organizational tools were available to them. Across the United States there was an increase in the late 1960s and early 1970s in local environmental organizing. This increase was due to a confluence of forces, including new developments in legal philosophy and federal law, reverberations of the social activism of the 1960s, the continuing power of the rhetoric of citizenship, and support from preexisting organizations including the League of Women Voters. This phenomenon is best described as "citizen environmentalism" as a nod to its decentralized, local nature and the prominent rhetoric of citizenship and participatory democracy.

Choosing Citizenship

The concept of citizenship was the most powerful of the identities available to the modern environmentalists. In the late 1960s and throughout the 1970s, new environmental groups across the U.S. chose names that included the word *citizen*: from the Citizen's Clean Air Committee to the Citizens Against Pollution and the Citizens Council for Clean Air, these organizations took special pains to identify themselves in a manner that laid claim to an individual's rights in a democratic society. The list seems endless: Citizens for a Better Environment, Citizens Clearinghouse for Hazardous Waste, Citizens Opposing Pollution, North Area Citizen's Conference, Tennessee Citizens for Wilderness Planning; nearly eight hundred nonprofit groups formed across the United States in 1968 alone had the word *citizen* in their name, with a large chunk of these being environmental groups.[3] Journalists across the nation picked up on the language, with headlines presenting "A Citizen's View of Air Pollution," declaring that "Citizens Demand" cleanup, and when change occurred, asserting that "Citizens' Pressure Results in Decree."[4] Activists in Pittsburgh nearly exclusively used the language of citizenship, referring to themselves not as environmentalists but as a "citizen's action group" or a "non-profit citizen's organization," and declaring that "citizens have to see that the laws are obeyed" in a newspaper story headlined "Citizens Wising Up." Countless government agencies picked up on the language at the same time: the Pennsylvania Department of Environmental Resources created a "Citizens Advisory Council" in the early 1970s, and the federal Environmental Protection Agency instituted a "Citizen's Advisory Committee on Environmental Quality" by 1971.[5]

In America in the 1970s, occurrences of the word *citizen* accompanied the oft-noted explosion in the use of the title *environmentalist* in public discourse. Even if *citizen* couldn't entirely keep pace with the increasingly popular *environmentalist*, the term generally appears first in describing activism, and just as prominently. A search of postwar newspapers from small towns across the United States finds that *citizen* appeared in approximately one-third of all articles that also used *environmentalist*, even as use of the latter word expanded. In the decade of the 1950s, 52 of 77 newspaper articles that used *environmentalist* also included *citizen*; this number rose to 83 of 234 articles in the 1960s and an incredible 25,980 of 83,552 in the 1970s. In a sample of 307 headlines from *Pittsburgh Press* articles and letters to the editor from the 1950s through the 1980s, the word *citizen* appears exactly as many times as the word *environment* or any form thereof; and in the text of those 307 articles, *citizen* far outstrips *environmentalist*, appearing first and more often. Even the word *public* appears in more articles than *environmentalist*.[6]

This sampling of public language implies that the concept of citizenship had important rhetorical, political, and legal value for the environmental movement, an importance that becomes more noticeable when we examine small-scale, local environmental activism. These small groups often included the word *citizen* in their names or in descriptions of their demography and actions. They laid claim to the right to speak in public and in governmental hearings by emphasizing their rights as citizens in a democracy. They petitioned for access to courts and to the inner workings of government using the newly evolved rationale of "citizen standing." Laying claim to "citizenship" in the early days of the modern environmental movement linked those activists with the language of "people power" and participatory democracy so prevalent in the last half of the 1960s and the first half of the 1970s, culminating in the reappropriation of revolutionary language surrounding the Bicentennial in 1976. The theme of citizenship appears so often in the rhetoric of these organizations that it is as important to their self-image as the language and philosophy of environmentalism.

It's even possible that environmentalists were uniquely attracted to this language: environmental groups were twice as likely as all other social and civic nonprofit groups formed throughout the twentieth century to include the word *citizen* in their name. Out of the nearly 12,000 environmental groups to apply for tax-exempt status from the Internal Revenue Service in the twentieth century, 179 had the word *citizen* somewhere in their name, and a further 242 featured the word *community*, 50 *public*, 29 *neighborhood*, and 664 *friend*. While environmental groups were more likely to include *land* (1,127), *water* (704), *environment* (656), and *forest* (316) in their names than *citizen*, this is to be expected—after all, land, water, environment, and forest are the reasons these groups existed. But it might be less expected to see that groups chose to include the word *citizen* more often than *earth* (147), *wildlife* (124), *air* (67), or *pollution* (40) as markers of their identity. These environmentalists were strangely attracted to the idea of citizenship; so much so that the language of citizenship, democracy, and people power rivals environmental themes in the first blush of activism in the modern environmental movement (see Chart 1.1).[7]

Environmental groups likewise came to choose the word *citizen* as a component of their names only in the 1960s; while no groups applying to the IRS had that word in their name before October 1964, more groups chose names that included that word in every decade after that point. The fact that environmental groups are far fewer in number than all other social, civic, and charitable organizations increases the significance of the observation that environmental groups were twice as likely as all other groups to include the word *citizen* in their names; out of all types of fund-raising charitable organizations—arts groups, public theaters, food charities, veterans groups, parent-teacher associations, civic boosters—environmentalists were more likely to chose to associate themselves with the concept of citizenship.

What did these early environmentalists mean when they called themselves *citizens*? They were of course laying rhetorical claim to the right to petition government, a right of citizenship enshrined in the Bill of Rights, declaring that they, as citizens, "belonged" in public discourse. But at the same time, they were defining citizenship in a specific way, rejecting definitions that emphasized voting and partisanship. For example, they did not describe themselves as *voters*, nor did they

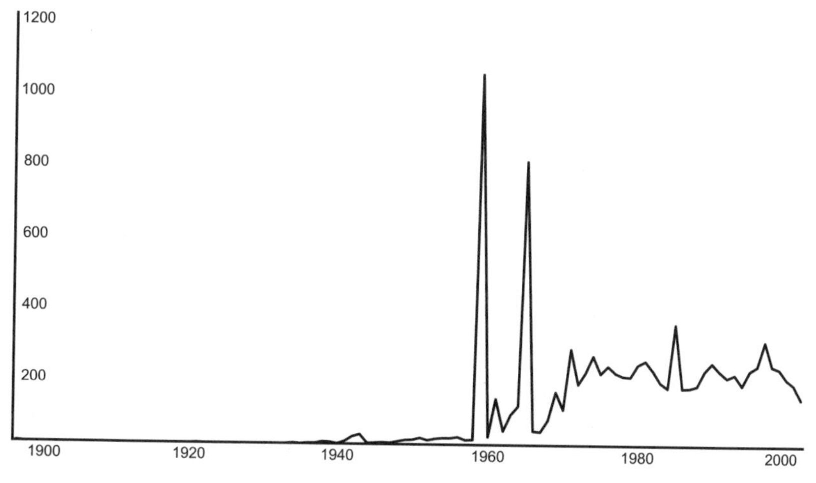

Chart 1.1. Groups with *citizen* in their name, by founding date.

endorse candidates or political parties. They were directly appealing to government through means other than the ballot box. Those who chose to describe themselves as *citizens* were casting those with whom they might not agree—*bureaucrats* and *industrialists*, perhaps—as being *not* citizens, perhaps even unpatriotic. They were also choosing the primary identity of *citizen* over *environmentalist*, possibly indicating that for them, *citizen* contained more populist rhetorical power. It implies that democratic participation shouldn't require professional expertise or industry clout, as one radio documentary put it: "most of those testifying were just plain, ordinary citizens; people tired of breathing dirty air." For these early environmental activists, the word *citizen* implied the same legitimacy and impression of democratic representations of the heartland that the word *grassroots* enjoys today.[8]

A further twist in the definition of citizenship was demonstrated at various public hearings across Pennsylvania in 1972 when Mrs. Joan Hays repeatedly introduced herself as a "professional citizen," causing the chair of one particular meeting to ask "what is the difference between a professional citizen and a citizen?" Hays had an answer: "A professional citizen is one who works at it full time and studies how to be one."[9] For Hays, citizenship meant active involvement, and her insistence on labeling herself a "professional citizen" echoed the countless other activists across the nation who used the idea of citizenship as a means of influencing government. This emphasis on citizenship should remind historians that the environmental movement was a late addition to the "rights revolution" of postwar America: a sweeping redefinition of citizenship to mean those individuals who were an active component of government instead of merely those who consented to be governed.[10]

The modern environmental movement is historically significant because of this redefinition of citizenship, itself a component of the social revolutions of the long 1960s, not because of any new developments in the environmental message or philosophy. Indeed, many if not all of the concerns of the 1970s environmental movement, from overpopulation to resource management to pollution and toxics, had been issues for decades if not centuries before. What made the movement significantly different were the newly substantive opportunities for public involvement in policymaking. The environmentalists of the early 1970s were not simply advising voluntary cooperation, as the Progressive reformers before them, nor were they just writing articles and books to focus public attention on environmental concerns, as had the activists of the 1950s and early 1960s. Rather, they were finding ways to compel mandatory action through courts and government agencies by laying claim to the right of substantive inclusion in policy matters.

In the case of the citizen environmentalists of Pittsburgh, substantive inclusion meant not merely advising, lobbying, or agitating, but rather taking an active, independent, and empowered role in the enforcement process.[11] Fully understanding environmental activism requires paying special attention to the messy and convoluted details of the enforcement process and the roles played by newly empowered citizens groups therein. Whether it's described as implementation, enforcement, or the day-to-day activities of pollution control bodies, this stage in policy development is insufficiently studied by environmental histories, which often spend more time on the public debates surrounding policy formulation. But enforcement—the application of policy decisions to real-world events—is the very definition of *substantive* involvement: if public involvement in policymaking did not change enforcement, it did not have any meaningful effect. The goal for activists was to gain the right, or standing, to substantially involve themselves in policy implementation and enforcement activities, and thus accomplish meaningful change.

Citizen Standing

The citizen environmentalists of Pittsburgh used a variety of tools at their disposal—their social capital or access to professional, religious, women's, and civic organizational networks; their status as professionals and women; the rhetoric of pluralistic democracy; the rhetoric of women's care for the urban environment; their relationships with government agencies—to establish their *standing* in both the court of law and the court of public opinion. As a legal term, *standing* refers to a plaintiff's right to bring a legal claim before the court. Throughout the late 1960s and 1970s, the courts extended the legal definition of standing. Deviating from traditional insistence that the parties have some economic interest in the matter at hand, new rulings included citizens who were not financially impacted by a wrong such as pollution. These new plaintiffs won the right from the courts to intercede in legal affairs based on moral, medical, aesthetic, or political arguments. This new legal doctrine was known as *citizen standing*, and the environmental movement was built on its foundations.

Standing is more broadly used here to indicate a citizen organization's ability to involve itself in public, political, or governmental affairs. Individuals and organizations attempted to prove or defend their standing in the courts and in the public sphere alike by demonstrating a number of claims: the impacts pollution had on their lives, their rights as citizens in a democracy, their numbers, their empowerment by federal legislation, their special status as wives and mothers, or their professional expertise. As such, the fight for standing was as much a rhetorical exercise as it was a change in legal philosophy, and it was a necessary step toward the greater goals of the citizen environmentalists.

In the courts, of course, the meaning of citizen standing is more tightly bounded. Standing itself refers to a plaintiff's right to bring suit, as opposed to jurisdiction, which describes that court's right to rule over the parties before it. As traditionally conceived, standing is determined before a plaintiff or petitioner can bring an action before the court, and is determined by that party's relationship to the respondent or status under the legislation in question. As one legal reference puts it, "A court may and properly should refuse to entertain an action at the insistence of one whose rights have not been invaded or infringed." One cannot appear before a court to ask for a ruling "unless one has in an individual or representative capacity some real interest in the cause of action, or a legal or equitable right, title, or interest in the subject matter of the controversy. This interest is generally spoken of as 'standing.'"[12]

But citizen standing depended upon a much wider reading of who might be an "interested party." The environmental movement had previously built upon this legal philosophy, especially in the 1965 case of *Scenic Hudson Preservation Conference v. Federal Power Commission*, which established that "aggrieved parties" might include "those who by their activities and conduct have exhibited a special interest" in the land or water in question. Later decisions expanded on *Scenic Hudson*, specifically dispensing with the legal requirement of land ownership or financial involvement as a proof of direct interest. In *Association of Data Processing Services v. Camp*, the Supreme Court wrote that "interest, at times, may reflect 'aesthetic, conservational, and recreational' as well as economic values." Perhaps the high tide of the citizen standing philosophy before the Supreme Court came in the 1973 decision in *United States et al. v. Students Challenging Regulatory Agency Procedures (SCRAP)*. There, Justice Potter Stewart wrote that in its previous decisions, the Court had "made it clear that standing was not confined to those who could show 'economic harm.'" But going even further in *SCRAP*, Justice Stewart wrote that "all persons who utilize the scenic resources of the country, and indeed all who breathe its air, could claim harm similar to that alleged by the environmental groups here," and since the *SCRAP* decision was in favor of the environmental group, claiming noneconomic harm could open the door to standing for nearly everyone.[13]

This expansive Supreme Court reading of who had the right to substantive involvement in legal matters encouraged citizen environmentalism on the local level all over the nation. After the battle over federal recognition of strict state air pollution standards in Montana, law professor J. P. McCrory urged the general public to become involved in legal proceedings on environmental matters. "It is clear that every one of us" could be adversely affected by pollution, said McCrory, "so we all potentially have standing" in the courts. At an air pollution workshop, he argued that all citizens could act as "private attorneys general" in demanding that the law be enforced. Montana was not the only state affected; in Michigan, new legislation placed every "citizen on a legal par with the state's attorney general in environmental cases," as *Time* magazine noted in 1970. The resulting law allowed "any private citizen [the right to] sue against a public nuisance on behalf of the general population, whether or not the nuisance effects him personally."[14] In Wisconsin, crusading law professor Joseph L. Sax made it his mission to legitimate the citizen suit and the public's right to represent its interests in court. Sax's efforts included both popular publications for a national audience and attempts to encourage citizens' suits in Wisconsin. In his 1971 book *Defending the Environment: A Strategy for Citizen Action*, Sax argues that courts were the proper venues for environmental activism. Rather than going through a regulatory middleman, citizens could directly access their government by bringing suit in a courtroom, and publicly airing differences and competing evidence. "Litigation is thus a means of access for the ordinary citizen to the process of governmental decision-making," Sax concluded. Environmental activists intent upon encouraging public intervention made Sax's book required reading.[15]

Encouraged by Sax and other popularizers, citizens increased the number and visibility of lawsuits based on the public trust doctrine. As a result of such pressure from below, the power of citizen suits marked a distinct break in the American regulatory framework: before 1970, standing doctrine "stood as virtually impregnable barriers against citizen-initiated cases," according to Sax. But new organizations exploiting citizen standing blossomed throughout the 1970s with the addition of new federal legislation requiring public input, provisions for citizen suits within federal acts, and increasing acceptance of the citizen standing doctrine. This would seem to be one of the bases for the phenomenon of the public interest research groups, or PIRGs. Starting with Ralph Nader's work as a consumer advocate, but branching out into dozens of state-level and campus-based groups, the PIRG movement created innumerable local activist groups focused on promoting public oversight of legally complex and technocratic environmental policy issues. Clean Water Action, for example, formed in 1971, promoted state and local advocacy groups on water quality.[16]

This generalized acceptance of citizen intervention in legal affairs relating to the environment is quite possibly the most significant legacy of the environmental movement. The ability to substantively intercede—to bring suit—was a validation for existing citizens groups and a reason to create new ones. While environmental historians have highlighted the national development of citizen standing or public interest suits with a special emphasis on Supreme Court decisions, it is also important to note the widespread popularity of citizen suits, and the regulations that encouraged them, across the nation. The inclusion of the public in policymaking, through the mechanism of citizen suits, defined a new period in the history of government and regulation in the United States; one in which we still live.

The Language of Participatory Democracy

Citizens groups were encouraged by the widely endorsed ideals of public involvement and participatory democracy in the 1960s and 1970s. The language and ideals of an active citizenry were popular across nearly all divisions of the increasingly fractious postwar America, and provided rhetorical and philosophical rationales for organizations attempting to gain access to the halls of power. The language of participatory democracy was virtually unassailable, which is why so many environmental organizations chose to identify themselves with citizenship. Armed with the patriotic language of the American Revolution, encouraged by cold-war exhortations for democratic expression, bolstered by the use of participatory democracy in the many social movements of the rights revolution, and quickly approaching a nationwide Bicentennial celebration, local environmental organizations rode a popular wave of support for their inclusion in the public sphere.

Much of the language used by citizen environmentalists came from the participatory democracy promoted by the social movements of the 1960s. The Port Huron Statement of the Students for a Democratic Society (SDS) spelled this out explicitly: "As a social system we seek the establishment of a democracy of individual participation, governed by two central aims: that the individual share in those social decisions determining the quality and direction of his life . . . [and that in] a participatory democracy, the political life would be based in several root principles [including] that decision-making of basic social consequence be carried on by public groupings." Political scientist and organizer Meta Mendel-Reyes, in her book *Reclaiming Democracy*, makes this argument explicit, using the words of Fannie Lou Hamer and the actions of the Mississippi Freedom Democratic Party to argue that "during the era of the civil rights movement and the New Left, groups of activists fought not only for the rights of all Americans to participate in politics, but also for a redefinition of the meaning of political participation. [Student Nonviolent Coordinating Committee (SNCC) and SDS] conducted the two most significant experiments in participatory democracy." Historian Robert Fisher agrees, noting that "in SNCC and SDS participatory democracy became an ideology of both goals and methods, a vague vision of how to get people together, act together, and win power together." This emphasis on democracy in the 1960s contributed to an emphasis on local political organizing throughout the 1970s, what Fisher calls "the New Populism."[17]

Part of this new populism was an increase in the number and type of neighborhood or community political organizations, particularly those organized or inspired by Saul Alinsky. In 1970, *Time* magazine called Alinsky "The Prophet of Power to the People" and paraphrased his motto as "a sharing of power . . . is what democracy is all about." Alinskyite groups were distinct from civil rights organizing and more closely linked to the labor organizing tradition; the resulting neighborhood groups specialized in tight-knit organization that created new centers of political power in minority, ethnic, or previously ignored city ghettoes. These included ACORN (Association of Community Organizations for Reform Now, founded in 1970 and momentarily famous in the 2008 presidential elections and subsequent politics), but also the many other groups associated with Alinsky, from TWO (The Woodlawn Organization) to BUILD (Build Unity, Independence, Liberty and Dignity) and FIGHT (Freedom, Integration, God, Honor, Trust). Throughout the 1970s, groups like PICO (first Pacific Institute for Community Organization as found in 1972, then later People Improving Communities through Organizing) were characterized by acronym names, the language of participatory democracy, and the shared goals of organizing disempowered locales to gain access to political decision making. Similar naming patterns and emphases show up in non-Alinsky groups descended from the civil rights movement; the Reverend Jesse Jackson founded PUSH (People United to Save Humanity) in 1971, later becoming the Rainbow/PUSH Coalition.[18]

The language of participatory democracy, citizenship, and public involvement in government was perhaps most noticeable in cold-war descriptions of America intended for international audiences. An account of municipal air control in a United Nations publication noted that "the key to success came through co-operative community action." Even if pollution threats cropped up again, the author promised that "the citizens are prepared to continue their anti-pollution campaign with unrelenting vigor, secure in the knowledge that they can depend on community action." Even more explicitly aimed at the cold-war foe, a Voice of America radio documentary titled *A Breath of Fresh Air* described participatory democracy as an essential component of a successful society, and devoted an hour to promoting GASP as a successful example of citizen involvement. "The people power which worked in Pittsburgh can work anywhere in the world," the narrator promised, somewhat grandiosely. Back in the United States, GASP trumpeted these propaganda claims as a justification of their status representing American ideals: "Now we are to be publicized behind the Iron Curtain by [the U.S. Information Agency] in *Amerika Illustrated*," said activist Michelle Madoff in a speech promoting GASP, "to show how a citizens' group works in a democracy."[19]

People-power rhetoric showed up in professional and business publications as well. In 1962, a breathless pamphlet from the National Association of Manufacturers, a lobbying organization for heavy industry in the United States, praised the inclusion of local citizens in air pollution reform efforts, arguing in italics that "*the surest and most effective way to combat air pollution is through cooperative community action.*" In the pages of the *American Journal of Public Health* in 1969, one author recognized the power of the "idealized Greek polis, . . . [the] New England town meeting, . . . [and] citizen participation in modern metropolitan government" in the minds of Americans, as a call for an increase in responsible and professional citizen involvement. While these exhortations for citizen involvement came from established, mainstream sources, how far away were they from ACORN's motto, "The People Shall Rule"?[20]

The language of citizenship and participatory democracy reached a rhetorical pinnacle in the year leading up to the Bicentennial of the American Revolution. Throughout the first half of the decade, local, state, and federal institutions had linked examples of successful public activism to the democratic ideals of the Republic. For example, the U.S. Department of Housing and Urban Development and the "American Revolution Bicentennial Administration . . . selected our Bicentennial as an opportunity to highlight the activities of 200 problem-solving community groups from throughout the country," and chose to publicize the actions of the new wave of citizens groups in a pamphlet titled, hyperbolically, "Horizons on Display: Part of the Continuing American Revolution." The completed pamphlet, printed by the Government Printing Office and promoted to various United Nations conferences and cultural exchanges, praised Pittsburgh's GASP "for its involvement and work in providing and maintaining citizen thrust and interest in continually improving air quality" and "organized community leadership and enthusiastic citizen involvement." GASP was presented with a Bicentennial flag to display for the year as a sign of their embodiment of the ideals of the revolution.[21]

While the many varied celebrations of the Bicentennial were, at best, ambivalent in the dispiriting era of the Vietnam War, Watergate, Altamont, and stagflation, they at least served to highlight the language of *revolution, citizenship,* and the importance of *the people* in public discourse. Whatever the "Spirit of '76" meant to Americans in the early 1970s, it reintroduced the language of democracy, ripe for reappropriation by citizen activists. Bostonians were urged to attend a 1973 Tea Party reenactment with the cry "Citizens of Boston, be prepared to make history!" The federal organizers of the Bicentennial subtitled their final report *A Final Report to the People,* while critics of commercialization of the Bicentennial founded the "People's Bicentennial Commission" and published their own book subtitled *A Planning and Activity Guide for Citizens' Participation during the Bicentennial Years.* In perhaps the most important action promoting participatory democracy, the Bicentennial taught an entire generation of Americans basic civics: the popular *Schoolhouse Rock!* animated shorts introduced the history of the Revolution, westward settlement, checks and balances, antimonarchialism, women's suffrage, and bicameral legislative schemes to the world of Saturday morning cartoons. "I'm Just a Bill," "No More Kings," and "The Shot Heard 'Round the World" all debuted in 1975 and 1976.[22]

The Bicentennial served to promote the importance of citizenship and participatory democracy, but ironically, historian Christopher Capozzola writes that early Nixon administration organizing for the Bicentennial celebration was met with "popular demand for greater participation in event planning." In response to these demands, as well as the pressures of the crumbling national economy and the mounting disaster of the Nixon presidency, the Bicentennial was turned over "to local governments and voluntary associations," and celebrated at the local level.[23]

Public Hearings and Dialectical Relationships

While citizenship and women's involvement supported environmental activism, interaction between federal legislation and local action translated that activism into meaningful transformation. In Pittsburgh and across the nation, local environmental activism had a dialectical relationship with environmental policy in the 1970s. In other words, the pervasive mood of participatory democracy and increased citizen participation associated with the civil rights, community organizing, antiwar, and women's movements created the political basis for federal legislation that specifically included citizen involvement in environmental policymaking. That type of legislation in turn created reasons for new organizations to form and for old organizations to become more active. Once functional, those organizations spent at least some of their time advocating for the legislation of further opportunities for public involvement at all levels of government. The opposing forces became intertwined in a cyclical and transformative process dependent upon the continuation of their differences: government agencies and politicians used public participation as evidence of broad public support for their goals, while citizens used the imprimatur of government-legislated public involvement as an organizational raison d'être and a rationale for increased involvement.[24]

Along with other causes, there also appeared to be considerable dissatisfaction with previous models of cooperative policymaking, necessitating the sort of break with the past embodied in the surprising public hearing described at the beginning of this chapter. The citizens seeking access to decision making repeatedly criticized the cooperative past. As the mayor of the borough of Springdale said in a 1972 public hearing in Pennsylvania, "the closed door brand of dedicated, sincere and positive pollution accomplishments always produce the same results, more pollution."[25] The root problem to be addressed here was not pollution—it was the political system that allowed the pollution to be created and blocked real solutions. As environmentalists, these citizens set their sights on institutional change as a means of environmental improvement.

Opportunities for this involved citizenry came directly from legislative requirements in a variety of federal programs of the late 1960s and early 1970s. From the Community Action Program of Lyndon Johnson's Great Society to increased participation in the decision making of the Army Corps of Engineers, the federal government altered the rules of public involvement. The 1967 Air Quality Act, the 1970 Clean Air Act, and the 1970 National Environmental Policy Act all included new and surprisingly specific requirements for public involvement in decision making, with the environmental impact statement mechanisms of NEPA being perhaps the most lasting and substantive development. In all of these examples, policy created participation—that is, choices made by government created opportunities, rights, and roles for citizen involvement. When the public rose to fill those roles, and even to press for further powers, the result was a measurable increase in both the reality and the rhetoric of public participation. Since this increase in public involvement in the 1970s, structured public involvement in environmental regulation and enforcement has become the accepted model for policymaking in the United States, as political scientist Richard Munton has noted: "Outside the corporate domain it is, today, quite difficult to find examples of environmental decision making where there has been no public consultation or other form of public involvement in the process."[26] This development was not inevitable; the decision-making model that Munton describes has its roots in the citizen environmentalists of the 1970s.

Previously, historians have offered the evidence of increased public interest in environmental matters as an explanation for the inception of the modern environmental movement—but there was an increase in all such groups, not just environmental organizations. Analysis of the increase in the number of environmental groups thus points to structural and political changes, not just a cultural shift that highlighted environmental concerns. These groups were created, at least in part, because of new legal rights offered to them and because of new legislative requirements for their involvement in policy matters. Empowered by federal legislation, citizens organizations used their newfound rights to influence locally unwanted land uses, forestry and agricultural issues, national park status, and toxic pollution—all through the power of lawsuits, publicity, and single-issue political organizations. At least three different forces fostered the creation and institutionalization of these citizens groups—requirements for public involvement in a wide variety of new legislation, the development of the "citizen standing" legal doctrine, and the language of participatory democracy.

This transformation in the relationship between citizen and government in the long 1960s marks an important divide in U.S. history. One historian has called this the "Silent New Deal" and a "second revolution," changing through legislation what the revolution in the streets could not. Others have offered the phrase "rights revolution" as a description of the legal and legislative transformation. A political scientist has described this moment at the division between the "New Deal Regulatory Regime," marked by cooperation and centralized expert control, and the "Pluralistic Regulatory Regime," characterized by confrontational debates, citizen activism, and specific policy tools to include, channel, and partially defuse competing interest groups. Historian Frank Uekoetter calls the previous era "the pseudocorporatist regulatory tradition" and dramatically declares not only that it "came to an abrupt and inglorious end in the late 1960s" but that new environmental policy "would leave behind the smoking ruins of the previous approaches without so much as a second look."[27] While these descriptions are useful, this new era of environmental policy might also be described as being marked by "citizen environmentalism," indicating that the increase in broad awareness of environmentalism depended upon a new inclusion of the citizen in the governing process.

The trend toward the federal government's inclusion of local citizens in the policymaking process began after World War II with housing and city planning programs. As Edward C. Banfield and James Q. Wilson noted in 1967, "Federal housing policy requires that before a city can receive federal funds for urban renewal projects, a local citizens' association must participate in and endorse the final plan. The law requires that this involvement include not only ready acceptance of the plans by the organized public but also active participation by the public in the planning activity." In this manner, federal grant programs intruded into local politics, raising the relative influence of citizens organizations by explicitly requiring their participation in decision making. The "organized public" described by Banfield and Wilson was empowered by the federal government to act in local affairs.[28]

Historian Robert Fisher argues that mid-1960s federal policy that specifically included provisions for public input was an attempt to mollify or mediate the rising importance of New Left and civil rights organizing from the first half of that decade. For Fisher, the loose and antiauthoritarian organizing structure of both SNCC and SDS became a model for 1960s community organizing, emphasizing participatory democracy over strong leadership. The Southern Christian Leadership Conference (SCLC), of course, had been involved in community organizing work related to the civil rights movement in the rural and urban South since 1962. One SDS community project, the Economic Research and Action Project or ERAP, promoted citizen participation and community organizing, particularly in poor neighborhoods of northern cities. Though very different in membership and strategy, SNCC, SCLC, SDS, and ERAP each gained national political prominence through active community organizing.[29]

The civil rights movement was thus the origin of citizen activism in postwar America, creating political pressure for the inclusion of the citizenry, providing organizational models of community organizing, and offering examples of successful public influence upon policymaking. For historian Robert Gottlieb, "the civil rights movement . . . shaped the protest movements of the decade," inspiring the New Left and the student movement, which eventually turned to community organizing. In fact, for most observers of the late twentieth century United States, the civil rights movements inspired all of the movements that followed.[30]

The federal response to the type of community organizing pioneered in the civil rights movement included programs that specifically required citizen involvement. Among these, programs targeting urban centers preceded many environmental policies that would eventually require public involvement. As a part of what would come to be known as President Lyndon Baines Johnson's "Great Society" programs, the Economic Opportunity Act of 1964 created a number of federal programs that promoted, funded, and legitimated local community activism in the interest of improving social conditions. The Community Action Program, a part of the 1964 Act, established more than one thousand federally funded Community Action Agencies in cities across the nation. The original legislation declared that the agencies were to be "developed, conducted, and administered with the maximum feasible participation of residents of the areas and members of the groups served." As a federal response to the civil rights movement, CAP attempted to empower local citizens organizations with federal funds and support. Though results from the

CAP and CAAs were arguably limited, Robert Fisher still thinks that the precedent was significant: "To its credit, the federal government . . . did support and fund a grassroots citizen action program." Perhaps even more significantly than CAP, VISTA (Volunteers in Service to America), the domestic equivalent of the Peace Corps, was also a part of the EOA of 1964. VISTA trained and placed volunteers in locally oriented community organizations across the United States, but focused on urban centers.[31]

While CAP and VISTA offer early examples of federally required citizen involvement, the model of decision making they provided was adopted in many different environmental policies scattered among agencies and their controlling legislation. In 1967 the Air Quality Act (AQA) provided that "all interested persons shall be given an opportunity to be heard" at conferences related to air pollution problems. Throughout the late 1960s, the Army Corps of Engineers redesigned many of its decision-making processes to specifically include public involvement. NEPA directed that, by the first month of 1970, all federal action had to be preceded by an "Environmental Impact Statement" (EIS) which was to be made available for public review and comment. The Clean Air Act Amendments of 1970 directed the states to develop and adopt implementation plans "after reasonable notice and public hearings." The Water Pollution Control Act Amendments of 1972 also included specific provision for citizen involvement, noting that "public participation in the development, revision, and enforcement of any regulation, standard, effluent limitation, plan, or program established by the [EPA] Administrator or any state shall be provided for, encouraged, and assisted by the Administrator and the States." While scattered across a number of new programs and policies, these requirements for "active participation" on the local level worked together to create a dialectical relationship between the federal government and community activists, wherein each encouraged the other.[32]

The dialectical process can be seen again and again in instances where the federal government encouraged public involvement. Both the 1967 AQA and the 1970 Clean Air Act depended upon state-level public support from environmentalist organizations, both to pressure and lobby state politicians and to legitimize what might otherwise be viewed as heavy-handed federal intervention. National support for the AQA came from a barnstorming state-by-state tour by Senator Edmund Muskie (D-Maine). Muskie was a "legislative entrepreneur" who carved out his own political power within the newly created Subcommittee on Air and Water Pollution by encouraging public involvement and then benefited from the resulting public outcry.[33] Public Health Service (PHS) scientists and administrators used similar tactics in the years leading up to the 1967 AQA, encouraging citizen action and then benefiting from the resulting public outcry to press for greater regulatory power and enforcement. As historian Shawn Bernstein has noted, "the expansion of the political community was one of the principal means of producing change [for the PHS]. . . . [P]rogram leaders relied upon the systematic use of public relations in efforts to shape the mass political support which would insure issue development and program growth. A pattern of public education, aroused public concern, and subsequent political pressure for increased federal air pollution control activities became a central influence on federal policy making."[34] Muskie and the PHS demonstrate the government's half of a dialectical relationship; success for both depended upon exhorting and then exploiting public pressure.

The citizen's half of the relationship also benefited. Muskie's subcommittee hearings on the proposed 1967 AQA were held in cities across the nation, including smog-plagued Los Angeles, and served to heighten national interest in air pollution issues and to encourage citizens to testify before federal hearings on local environmental matters. Muskie's actions in relation to the 1967 AQA illustrate a change in the type of activism empowered by federal law. In Los Angeles County throughout the 1950s, the local Motor Vehicle Control Board had "actively sought industry cooperation, talked of teamwork and a division of labor within the industry, let auto company laboratories do its testing, and occasionally enjoyed the free technical advice" of Detroit automakers, according to historian Scott Hamilton Dewey. But when Senator Muskie traveled to Los Angeles, the public hearings for the 1967 AQA seemed to offer encouragement and validation for newly critical citizens groups. These hearings marked a departure in Los Angeles. While previous organizations such as Stamp Out Smog (SOS) were largely supportive of local air pollution control efforts, after Muskie's 1967 public hearings a new organization known as Group Against Smog Pollution (not to be confused with Pittsburgh's GASP) criticized local efforts more directly. Along with yet another group known as the Clean Air Council, these groups turned against the Air Pollution Control District, loudly criticizing its efforts throughout the early 1970s. Here, increased public input created an adversarial relationship; now that there were more voices being heard, some of them inevitably were not on the same side as the government.[35]

In response to this push for increased involvement, the 1970 Clean Air Act Amendments (often referred to as the 1970 Clean Air Act or CAA) built upon the foundation of the 1967 AQA to create a novel method for the federal EPA to enlist the states in the process of controlling their own pollution.[36] In the 1970 CAA, Congress required the states themselves to come up with their own air pollution standards, to hold public hearings on standards and plans to meet them, and finally to send the plans to the EPA for review. The State Implementation Plans (or SIPs) were required to meet federal minimum standards for criteria pollutants, to present a workable plan for state regulation and enforcement, and most urgently, to be submitted to the EPA by January 30, 1972. This system allowed the EPA unprecedented control, in two specific ways: First, the federal agency would have the authority to quickly specify National Ambient Air Quality Standards for a number of pollutants. Second, the EPA would be able to compel the states to quickly write and defend their own plans to meet those standards, without having to face the unpleasant political ramifications of "forcing" strict regulation upon the states. Instead, the SIPs would force the states to regulate themselves.[37]

The SIP process embedded in the CAA was a remarkable opportunity for citizen involvement. A publication of EPA's predecessor the National Air Pollution Control Administration (NAPCA) from 1970 emphasized the new political opportunities for community organizations by noting that "the Clean Air Act provides opportunities for all who are concerned with air pollution to participate in the decisions being made about the quality of the air we must all breathe. These decisions will more often reflect community needs and values if the public will take advantage of these opportunities, a fact repeatedly being established throughout the nation."[38]

In response to these sorts of exhortations for involvement, more than seventy different citizens organizations wrote to the EPA's Office of Air Programs (OAP) with questions or comments related to the SIP public hearing process in 1971 and 1972 alone. As did many others, a representative of the Northside Environmental Action Committee of Indianapolis, Indiana, wrote to complain about the speed of the SIP process. Under federal requirements, the plans were to be written, presented to the public for a comment period, rewritten, and submitted to the EPA in just nineteen months. Stanley H. Vegors, Jr., wrote, "on behalf of the Citizens' Environmental Council of Pocatello, Idaho, concerning the regulation proposed by the state of Idaho pertaining to air pollution," and recommending that the EPA deny the state's SIP based on "the specific criticisms which we raised concerning these regulations." From El Paso to Indianapolis to Louisville, organizations wrote to the EPA administrators to entreat or encourage EPA intervention in state matters. William Darrah, director of a Hawaii-based organization named Life of the Land, wrote to EPA Director William Ruckelshaus in 1971 to argue that more strict federal standards should supersede looser state regulations, and threatened legal action under the 1970 CAA if the EPA failed to act: "in the event that EPA's position is divergent from ours please consider this letter as formal notification . . . that it is our intent to commence legal proceedings 60 days from this date in the District Court." Conversely, where it was in their interests, some citizens wrote to the EPA to argue that more strict state standards should supersede looser federal standards, as in the case of William Tomlinson of the University of Montana Student Environmental Research Center.[39]

Federal agencies did not just accept citizen input; they also actively encouraged it, promoting public intervention in pamphlets, speeches, and programs. In 1970, the NAPCA produce a nationally distributed pamphlet to encourage citizen involvement in all levels of environmental policymaking. "Community Action for Environmental Quality" used plain language to prepare citizens groups for what was expected to be a decade of environmental change. New federal and state environmental standards were expected, and "citizen groups can help see to it these standards are good and stiff," noted the text. It went on to identify public hearings as an important opportunity for groups to take advantage of newly legislated opportunities for involvement: "The law requires that the states hold public hearings and give all interested parties a chance to comment on the proposed standards. Citizen groups should make it their business to find out when the hearings will be held, and prepare for them." Using language remarkable considering the source, the pamphlet went on to paint industrial representatives as the villain in public hearings, warning that they would urge slow and methodical analysis as a means of evading enforcement and preventing citizen involvement. "To counter this kind of preventive participation citizen groups must present the case for effective control in force, and with knowledge of the specific issues involved." The pamphlet's authors recommended that after the standards were set citizens might need to form a watchdog group and be ready to fight in court to oversee enforcement.[40]

Once established, the EPA set out to further educate and mobilize the very organizations that were already lobbying the agency. Like the NAPCA before it, the EPA promoted further citizen involvement with examples of model organizing. The agency produced a pamphlet in 1972 titled "Citizen Action Can Get Results." After discussing examples of public involvement from Chicago, Pittsburgh, and San Francisco, the pamphlet's authors concluded that "citizen involvement and action has become an integral part of the movement for environmental quality in the United States."[41] To further that movement, the new agency proposed an equally novel presidential commission, the Citizen's Advisory Committee. Internal correspondence shows that EPA administrators instructed the group to think of itself not as an advisor to the president, but "as a vehicle by which citizens and citizen groups throughout the country can be acquainted with the implications of environmental problems and can be equipped to participate meaningfully in action programs."[42] EPA administrator William Ruckelshaus met with representatives of the CAC in order to focus their attention on preparing citizens groups across the nation for their role in creating SIPs. The EPA briefing memo for that meeting noted that "the 1970 Clean Air Act Amendments require that there be a public hearing held in each state prior to the submission of an implementation plan. . . . Citizens need to be appraised of what role they can and should play at these public hearings, how they should relate to State and local control agencies in the course of plan development, how they can be better informed on the problems of industry" in responding to potential regulations.[43]

Across the nation, environmental activists seized the opportunity for public involvement afforded by the 1967 AQA, and spoke at length about encouraging additional citizen activism. When federal legislation prompted the creation of a new method of regulating local air pollution in Washington State, organized groups of environmental activists were an integral part of the policymaking process. Citizen groups attended hearings in 1967 to support state legislation creating a new local control agency; were an active part of SIP hearings to meet EPA requirements in 1970; and were again active in hearings as a part of the 1977 CAA amendments. In all of these instances, citizen involvement in policy and enforcement matters was prompted by federal regulations requiring public participation in state air pollution control.

In 1967, King County officials announced that Seattle, Washington, had been chosen to host the headquarters of a new, three-county air pollution control agency.[44] The Puget Sound Air Pollution Control Agency (or PSAPCA) was created by that year's revisions to the Washington State Clean Air Act, and funded by local and state taxes as well as matching funds from the NAPCA. The PSAPCA wrote new, multicounty air pollution control regulations and followed a procedure requiring emission sources to formally petition for the right to operate under PSAPCA regulations or, in special cases, to operate in violation of those regulations.[45]

The entire PSAPCA Board of Directors heard permit or variance requests through the 1970s before a public audience made up of citizens groups, which often prompted or supported enforcement action. A retired PSAPCA environmental engineer, James Pearson, wrote in his memoir that "we were well-acquainted with many groups of citizen activists who were most vocal in their opposition to air pollution." Pearson notes that the groups had several characteristics in common: "they all managed to come up with catchy names and acronyms; [and] they never had as large a membership as they claimed since they had dedicated leaders who did most of the work."[46] While Pearson is supportive of some of the local ad hoc groups, and of Seattle chapters of national organizations, his memoir is generally quite critical of citizen involvement. Most of the groups he dealt with "absorbed a lot of our Agency's time" or "they were a pain in the ear." For example, "A group calling itself 'COPE' (Committee on Polluted Environment) . . . got a lot of undeserved attention." Pearson's somewhat grumpy assessment of citizen involvement is hardly the first nor the last time that such a thought has been expressed by government officials. But Pearson's statement demonstrates an irony of the dialectical process: the very citizen groups Pearson criticizes actually lobbied to help create the organization that employed him.[47]

Like events in Washington State, air pollution debates in Montana illustrate the dialectical relationship between federal and local power. Events surrounding the creation of Montana's SIP demonstrate how the EPA's method for creating SIPs empowered and encouraged the formation of local citizens groups, often in opposition to state or city politicians or enforcement agencies. In practice, the citizens groups would prove a surprisingly powerful participant in Montana's development of new air pollution policy.

Montana's first state air pollution law was passed in 1967, with support from citizens organizations, conservation groups, and newspapers in cities laboring under pollution from extractive industry.[48] Those forces again became active in 1970–71 with the announcement of public hearings for the creation of the first SIP to meet the deadline established by the EPA. As a result of a vast public outcry and extensive participation in the public hearings, the Montana State Board of Health wrote a SIP that was substantially stronger than EPA's ambient air quality standards. In 1972 Montana's outgoing governor, Forrest H. Anderson, refused to forward the SIP to the EPA for approval, effectively vetoing it until the seven-member state Board of Health submitted it over his objections.[49]

Across the state, the very citizens groups that had been created to speak out at the original SIP meetings rose to defend the plan against Anderson. One of the most prominent of these groups was the Gals Against Smog and Pollution, known in Montana simply as GASP but referred to here as Montana GASP. The "gals" of Montana GASP were a group of educated, middle-class women associated with the University of Montana in Missoula, and organized in part with the assistance of the League of Women Voters.[50] Marilyn Templeton of Montana GASP "outlined the proper methods of preparing to testify before the board of health," and argued that citizens "have to be the watchdogs—this is what GASP is trying to do."[51]

What is notable about Montana's quarrel over air regulation is that federal intervention facilitated the participation of the very citizens groups that supported the SIP and lobbied the EPA for its passage. Mr. and Mrs. John D. Grove urged Ruckelshaus to accept the state's stricter standards, noting that "we were involved in planning for a local workshop to inform and alert people to the forthcoming hearings on the state's implementation plans. This endeavor was financed by your office and we do want to thank you for a very excellent opportunity to educate people concerning our state air pollution problems. . . . The money was very well spent and we do hope this will be a continuing program."[52] The Groves' letter indicated that the EPA had encouraged the creation of the very citizens groups that were agitating in Montana to supersede EPA air pollution standards. This is the dialectical relationship at its most obvious: federal funding had fostered and trained local citizens who then agitated to provide public support and political momentum to allow the federal government to impose strict regulation upon the states.

All of this encouragement of citizen involvement led to an *expectation* of public oversight of environmental policy: in January of 1972, Albert Smith angrily confronted representatives of the Pennsylvania Department of Environmental Resources, excoriating them not only for the perceived weakness of their proposed environmental regulations but also for their actions excluding citizen involvement in the policymaking process. Speaking for the "Ad Hoc Emergency Committee for Clean Air"—which had been created only four days before and consisted of nine organizations, some of which had themselves only been recently formed—Smith alleged that "the citizens of Pennsylvania have been denied an effective voice in the implementation planning process by maneuvering within DER to accommodate industrial interests." In fact, said Smith, changes in the proposal "rendered much of the citizen testimony given at [previous] hearings inapplicable to the new proposals. The new proposals have received neither public notice nor comment. At least one citizen member of the Board itself was not informed of these relaxed proposals." Based on the exclusion of citizen testimony and lack of public hearings, Smith called for the proposed regulations to be shelved.[53]

While the SIP process was an important step in creating opportunities for public hearings, the section of NEPA with the greatest long-term impact on public involvement in policymaking was the requirement that every "report on proposals for legislation and other major Federal actions" include an assessment of that action's impact, good or ill, on the environment.[54] NEPA required that those reports, later known as Environmental Impact Statements, be written with oversight and comments from the public, the president, and an independent group of experts known as the Council on Environmental Quality. Additionally, any citizen could challenge the thoroughness and adequacy of an EIS, though not its conclusions, in a court of law. In effect, the EIS process created two entirely new venues for mandated public intervention in the environmental policymaking process: first, in the public comment stage, and second, in the courts after the completion of the EIS.

NEPA's requirement for EIS creation and review was an attempt not to mandate environmental values or restrictions on development, but rather to create a public space for policy debate. Political scientist Serge Taylor has argued that "in effect the law grandly but vaguely commands the agencies, 'Think more carefully before acting!'" In practice, the requirements for statements necessitated citizen involvement to demonstrate the political legitimacy of the federal regulation, but simultaneously fostered public opposition and legal challenges. As Samuel P. Hays notes, the EIS "provided a wedge for wider involvement in administrative choices." While the EIS process itself did not require government regulation favoring environmental protection, it "played an important role in opening up decision making to the environmental public."[55]

In a variety of ways, federal agencies were actively recruiting and empowering the citizens groups that would soon begin to criticize, lobby, and take legal action against polluters, local government, and even the federal agencies that had encouraged them in the first place. The widespread use of public hearings as a mechanism of policy formation is quite evident in the creation of the 1970 CAA and SIPs. But the two most long-lived mechanisms that promote public involvement in environmental policymaking are the EIS and citizen standing. Together, these administrative and legal developments served to solidify the power of what otherwise might have been a transitory public movement.[56]

The League of Women Voters and Public Hearings

Almost all documented examples of successful citizen movements for inclusion in decision making in the 1970s included, in some manner, the organizational and leadership skills of women, and the citizen environmentalists were no exception. Women, as Andrew Hurley puts it, were "the recognized guardians of domestic welfare, accept[ing] primary responsibility for the maintenance of suburban communities."[57] Wherever women became involved in local environmental organizations, they were often assisted and supported by local chapters of the League of Women Voters. Although the League itself was founded long before the 1960s, it served an important function as a promoter and facilitator of new activist organizations in that period. The League's longstanding emphasis on civic education and promoting citizenship meshed remarkably well with the groundswell of participatory democracy. With the help of the League, women not only drove the issues of air pollution control but spearheaded movements to reform regulation by including substantive public oversight in a variety of issue areas.[58]

In countless examples, the League functioned in dual roles: local chapters engaged environmental issues directly, and also worked to create and encourage additional citizens groups, thus vastly multiplying its impact. The NAPCA and its successor, the EPA, both monitored and encouraged League activism. One EPA field operative frequently commented on League activism: "My compliments to the League of Women Voters of Nevada for the best conference of this type that I have attended," she wrote to her superior. "The approximately 100 attendees represented all geographic and organizational diversities of the state." Later, she reported additional work by the League in a different state: "Another good workshop to add to the success list of the League of Women Voters. The approximately 70 attendees represented all geographic and organizational diversities in North Carolina. . . . For many of the out-of-state areas, this conference represented a first introduction to air pollution control." League women not only trained the environmentalists who spoke at the SIP hearings, they appeared themselves as representatives of the public: in El Paso, Texas, a League-sponsored SIP hearing featured both a history of air pollution in the region from Mrs. John Brient, president of the League of Women Voters of El Paso, and testimony from Mrs. L. H. "Bunny" Butler, of the League-trained Group Against Smog and Pollution.[59]

In Montana, the League trained citizen activists to speak at public hearings, and when the governor blocked the resulting legislation, intervened directly, lobbying the EPA for approval of the SIP.[60] In Pittsburgh, the League held training workshops in 1969 for public hearings that led to the creation of local groups, then cosponsored more training workshops with those local groups in order to train citizens to speak at more public hearings in 1971. In Indiana, the "most influential" of the civic groups concerned with environmental issues was the League, which popularized scientific knowledge, promoted public involvement in city government, and supported the creation of even more public groups.[61] In Los Angeles, journalists Chip Jacobs and William Kelly argue that in order to make meaningful inroads on smog control, local women and "the League of Women Voters concluded that the political apparatus needed to be shaken up and remade."[62] Throughout the state of Pennsylvania, the League involved itself in policymaking in as many ways as possible, directly writing letters to policymakers urging both increased environmental regulation and an increased ability for citizens to substantially involve themselves in public matters. Mrs. Eugene M. Landis, environmental quality chairman, wrote to recommend several specific alterations to state pollution law expanding the rights of the public, recommending that "sufficient advance notice of public hearings should be given for full citizen participation. . . . Lay persons should be named to state scientific committees to reflect the viewpoints of the communities in problem areas. . . . Court procedures should be set up on a state level protecting the right of the citizen to bring violators of air pollution regulations into state courts. The guidelines used should be those which allow citizen class-action suits on a federal level."[63] While these examples demonstrate the type of actions taken by the League, there are countless other instances in which the League of Women Voters weighed in on environmental matters, either directly or by facilitating the work of other groups.[64]

The League was significantly involved in preparing citizens for the public hearings required by the new federal environmental legislation described earlier in this chapter. In Pennsylvania, the EPA funded several organizations—including the League—to train, organize, and encourage citizens to present testimony at public hearings for SIPs. At the Pittsburgh training sessions, League representatives and members of environmental groups coached prospective speakers, recommending that individuals speak in person, bring extra copies of prepared material for the stenographer, and describe "a specific event which happened to you personally and which can clearly be tied to air pollution or present facts citing your source." For its part, the Allegheny County Council of the League presented workshop participants with the names and professional backgrounds of Environmental Quality Board (EQB) members to better prepare prospective speakers to appeal to the EQB. The League's efforts to prepare the public to participate in the air pollution hearings were actively supported by the EPA's Office of Public Affairs. The OPA funded "Community Air Implementation Workshops" across the nation, holding a general call for proposals throughout 1971 from interested sponsors. OPA mailing lists indicated 134 individuals and groups received funding for these workshops, including representatives from twenty-one states and the District of Columbia. Sponsors of the workshops came from a long list of public health, environmental, and conservationist organizations.[65]

In all, seven hearings were held in cities across Pennsylvania, and the League had a hand in all seven, boosting attendance and participation beyond expectations.[66] The EQB meeting in the state capital on December 2, 1971, drew 300 audience members and thirty-eight testimonials; of these, there were three representatives of League chapters, six from various community-level environmental organizations trained for the occasion by the League, four representatives of TB or respiratory health organizations, twelve industry representatives, one speaker from the American Association of University Women, and a number of private citizens. In Philadelphia, similar numbers of citizens spoke, and 350 audience members showed up. Minutes of the EQB noted the long days logged by EQB members at these hearings: meetings in Harrisburg and Philadelphia begin at 9 A.M. and "lasted until 7 P.M. In Pittsburgh, the hearings lasted until 11 P.M. on Friday night and continued on Saturday from 9 A.M. to 1 P.M. About 120 persons testified and there were about 600 attending the Friday session." Eventually, EQB members gave up counting speakers, and began measuring written statements in linear feet: "Mr. Sussman informed the Board that there is over two feet of testimony and studies submitted, which does not include the transcript."[67] Two and a half days of testimony at the state EQB meeting in Pittsburgh were filled by citizens coached and prepared for their roles by the League of Women Voters; their statements were extensively discussed in the local press, and so many showed up that a second hearing was scheduled for the following month.[68]

The crowds at the early 1970s public hearings are all the more significant because, exactly ten years before, the state had held similar public hearings on a previous change in air pollution control regulations, and almost no representatives of the public had appeared. At a September 21, 1961, public hearing in Clarion, Chair Victor Sussman admitted, somewhat dejectedly, that "We will open the hearing now, even though there is only a small group here." Only four statements were made at that hearing. Five days later in Wilkes-Barre, fifteen persons made statements—but thirteen of those were representatives of industry. Even more lop-sided was the hearing in Washington, Pennsylvania, where no citizens spoke about the proposed air pollution control regulations, but fifteen representatives from industry, unions, and local government spoke in opposition, including representatives from the Western Pennsylvania Coal Operators' Association, Pittsburgh Coal Company, United Mine Workers of America, U.S. Steel, Emerald Coal and Coke Company, Rochester & Pittsburgh Coal, Buckeye Coal, Allegheny Pittsburgh Coal, Mather Collieries, and Duquesne Light. The many other hearings on this 1961 regulation change were similar: representatives of citizens groups, or citizens themselves, made up less than 10 percent of the speakers at announced public hearings.[69]

By September 1969, the reverse was true. Public hearings in Pittsburgh overflowed with citizens, some outspoken in their right to be involved in decision making. Citizen John Tabor spoke at a hearing on the 9th: "I am a layman, not a chemist or a physicist, but I am very clear on one thing: We have one duty, and that is to protect the health and life of the people of Pittsburgh. If weakening standards for air pollution is going to jeopardize that health, I strongly oppose the weakening." The representative of the United Steelworkers had even adopted the language of citizenship, choosing that identity over his other credentials: "I do not come before you today as a doctor or engineer who has made it a lifetime project to study this most complex problem. I do come before you as a citizen of the great Commonwealth of Pennsylvania," said Julius Uehlein, director and secretary-treasurer of Pennsylvania's Legislative Committee of United Steelworkers of America. Other members of the public spoke without specifying any professional or political expertise to justify their inclusion in the hearing. Francis O'Toole spoke, and movingly announced that "I am a victim of emphysema, and by that I guarantee that this won't be a long speech because I don't have the air to do so." Dorothy Canary, a member of the Cambria County Citizens for Clean Air, announced that "I have been at countless meetings, and all I hear is the same old jazz every time, but I represent thousands and thousands of children in my area where I take film and pamphlets and all sorts of educational material to educate children to understand what air pollutions [*sic*] is all about because they don't understand it at all. . . . [Y]ou see little children going to clinics with bad lungs, coughing and bad eyes, and anything you can think of a child today is afflicted with." Two hundred fifty-six pages of citizen testimony fill the transcript, to the point that one speaker, Thomas A. Scott, begins to notice the lack of industry representation: "While I haven't been here all day, I haven't noticed while I have been here anyone in a policymaking capacity from any of the major industries in the area. There has been no member of any board of directors saying, We are more than willing to co-operate, and I am rather saddened by this and would like it entered on the record that, as a citizen, I am rather disappointed that industry has taken such a weak stand in the cause of air pollution, for certainly, if we are going to obtain their comments, where are they today? I don't see them."[70]

The attendance, language, and tone at public hearings on the same type of legislative matters in the late 1960s and early 1970s were very different than they had been in the early 1960s. Something had happened in the intervening years. Along with the increase in interest in environmental matters, cultural, legal, and legislative changes had encouraged and supported public intervention in policy matters to bring power to the public hearing. While there are multiple forces that encouraged increased citizen participation, it's also significant that almost all of the speakers who appeared in the Pittsburgh hearing discussed above were trained to do so by the League of Women Voters.

Citizens' rights and the importance of participatory democracy became important themes for those who were coached to speak by the League of Women Voters. As these hearings were ostensibly concerned with air pollution control, it is surprising that so much time at the hearings was occupied by discussions of public participation. For example, the competition for public support at the hearings between business, industrial interests, and environmental organizations was thrown into the spotlight. Environmental groups criticized the state Chamber of Commerce for blindly supporting industry, while business concerns decried the tumult of citizen involvement.[71] According to Pennsylvania's Environmental Hearing Board examiner and Citizen's Advisory Committee member Norman Childs, "One thing that came through loud and clear . . . is that citizens now want to have a voice in any proceedings that are going to affect their health and welfare in the future." Reporting on the public hearing, a Pittsburgh newspaper made an editorial observation about the new attitude on display: "Although it was not on the official agenda, another fact became evident at the hearings—a general attitude of distrust on the part of the public. Spectators and citizen witnesses alike were inclined to discount testimony of any industry witness no matter how many figures he cited."[72] Pittsburgh environmental activist Michelle Madoff used the opportunity of her January 27, 1972, statement to the public hearing not only to comment on the pollution requirements of the SIP but to recommend that the SIP include requirements for substantive inclusion of citizens in the resulting regulatory framework: "It is not good enough to hold public hearings on a plan, and then claim you have listened to the citizens' wishes and arbitrarily write plans *that exclude citizen participation as a part of the regulatory procedure*."[73] Madoff would not accept token or symbolic inclusion; such an opportunity would only be useful for her insomuch as it allowed an opportunity to argue for substantive inclusion. There are more examples: Mary Timney McCormick, speaking for the Pennsylvania Division of the American Association of University Women, argued that "the value of open hearings is self-evident. While it is expensive to hold them, nonetheless the people must have the opportunity to present evidence concerning an issue which affects them so personally." Dr. Richard Diehl noted that "the Clean Air Act of 1970 requires active citizen participation by law. The citizens have been given the right to review their state's implementation plan and actually should help write it." And Albert Smith, continuing the testimony excerpted previously in this chapter, called for even more allowance for citizen involvement: "The major omission from the permit section is that there is no provision for public participation. . . . [W]e do recommend that this section be amended so as to provide public announcement of all permit applications, denials, and grants, and that such information be available to the public upon request."[74] The dialectical process was at work, as newly empowered citizens used their access to public hearings to call for yet more citizen involvement.

• • •

While this chapter argues that legal, legislative, organizational, and cultural forces in the 1960s and 1970s brought "power to the public hearing," the question remains: what qualitative effect did this change have on the policymaking process? The fact that citizens groups existed, were partly created by legislative and legal rights, and capitalized on the language of citizenship and the zeitgeist of participatory democracy doesn't explain what these groups actually accomplished. For Pittsburgh historian Roy Lubove, the end of the cooperative model of local policymaking made the process of urban planning more difficult. He argues that the "events of the 1960s and 1970s—the civil rights movement, community action and neighborhood organization, growing disillusionment with large-scale urban renewal and its social dislocations, . . . complicated the decision-making process and reduced the relative power of the business-centered civic coalition." But for other observers, the end to the cooperative model and the rise of the citizen environmentalists marked the beginning of real progress. Theda Skocpol and Morris P. Fiorina write that "[m]any observers find such transformations heartening. In their view, the United States has moved away from a politics of narrow interest group maneuvers toward a more inclusive and pluralist debate about public good." Richard N. L. Andrews argues that the environmental movement's "most sweeping effects have been to broaden the scope of political debate . . . to increase public awareness of the possible consequences of government and corporate decisions, and to diversify access to political decision processes for citizens who may be adversely affected by them. In short, it has contributed significantly to the democratization and pluralization of American politics."[75]

Whatever the assessments of historians, small groups of activists had a significant impact on national policy. While such small organizations could never hope to garner the same journalistic attention as larger national environmental organizations, community-based groups with names like GASP, CAN, and COPE kept up a steady drumbeat of environmental activism surrounding local polluters, state regulators, and municipal politicians. With the assistance of federal legislation that promoted citizen involvement, these small organizations actually enjoyed a considerable cumulative impact nationwide, making up the groundswell of popular activism that supported a more visible national environmental movement.

It is clear that the transformation of environmental policy at the beginning of the 1970s was as much an institutional revolution as it was a philosophical one. The pluralistic model of policy formation was adopted at many levels of government, from the county-level variance board in Pittsburgh to the federal requirements for public hearings and citizen oversight that were a part of the CAA amendments and NEPA. Even if citizens groups could not always win in their various battles for environmental protection, they were often successful in becoming a more substantive part of policy formation, on the local and the national levels alike. From Washington, D.C., to Washington State, local citizens battled throughout the 1970s to become a more substantive part of the air pollution control process, aided by federal intervention, the organizing strength of women activists and the language of citizenship.

• • •

Back in the contentious ambush hearing of September 1969, the Pittsburghers who showed up for the health department hearing were, most likely, unconcerned with any broader significance for their actions. The central motivating factor for them was the Smoky City itself, the legendary Pittsburgh air that one could almost taste, frequently smell, and always see. Still, these citizens were somehow motivated and organized enough to come together and criticize not only the proposed air pollution policy but also the very process that had excluded citizens from making policy. Their environmental concerns may have been the motivating force bringing them together, but once they embarked on a journey toward their environmental goals, their focus turned toward the tools and strategies that could get them there. They became citizens first and environmentalists second, first demanding their rights as citizens in a participatory democracy in order to then work toward their environmental goals. In retrospect, perhaps the Bicentennial is as significant as Earth Day in the history of America's modern environmental movement.

2 The Smoky City

PUBLIC INVOLVEMENT IN CONTROLLING AIR POLLUTION IN PITTSBURGH

Pittsburgh town is a smoky ol' town, Pittsburgh.
Pittsburgh town is a smoky ol' town, Pittsburgh.
Pittsburgh town is a smoky ol' town,
Solid iron from McKeesport down.
Pittsburgh, Lord God, Pittsburgh.
. . .
All I do is cough and choke in Pittsburgh.
All I do is cough and choke in Pittsburgh.
All I do is cough and choke,
From the iron filings and the sulfur smoke.
In Pittsburgh, Lord God, Pittsburgh.
—Woody Guthrie, "Pittsburgh Town"

Pittsburgh's air pollution problem makes an excellent case study of public involvement in political matters, and it does so for two reasons. The first is illustrated in the lyrics to folksinger Woody Guthrie's "Pittsburgh Town": the political matter at issue was so visible, so omnipresent, and so fundamental to the city's image that it permeated almost everything in the city and in the minds of its residents alike. First and foremost in the minds of Americans, Pittsburgh was a "smoky ol' town." Air pollution is unique in this way; it disperses more easily and more widely than does water pollution, which flows only downstream. It circulates more quickly and broadly than does solid waste, which sits more or less where humans put it. Air pollution spreads out across the city and countryside, without regard for political boundaries and largely outside of human control, making for a remarkably difficult policy problem. The second reason might also be implicit in Guthrie's lyrics, written in the tradition of folk protest against forces seemingly arrayed against the interests of the common people. For the vast majority of the city's history, political responses to smoke and air pollution in Pittsburgh excluded public representation. It was not until the late 1960s that there was substantive and meaningful inclusion of the public. The importance of the issue, combined with the transition in public involvement, leaves historians both with copious records and effective comparisons of control attempts with and without public involvement. Pittsburgh's dirty skies, and more than a century's worth of attempts to clear them, provide a microcosm of larger political themes. The case of Pittsburgh thus deserves some explication, and this chapter describes the history of both the forces that created the pollution problem and political attempts to resolve it.

Making the City Smoky

In Pittsburgh, a variety of forces combined to concentrate the effects of air pollution, including meteorology, geography, technology, human settlement, industrial development, and the available fuel. Throughout the nineteenth and early twentieth centuries these forces made Pittsburgh known, nationally and internationally, as a dirty city; the town was defined by its air pollution as much as it was by its industrial production. Pride in industry competed with an emerging, progressive vision of urban life as the negative impact of air pollution became increasingly evident and set Pittsburgh apart from the rest of the nation. That vision spurred a series of attempts to control the production or ameliorate the consequences of smoke, and later, air pollution.

The settlement that eventually became the city of Pittsburgh was founded on a narrow triangle of land where two rivers met to form a third. Deep in the forests of Southwest Pennsylvania, this natural formation was attractive to Native Americans and Anglo explorers alike, and an obvious location for trade, fortifications, and settlement. The confluence of rivers was a strategic hub for transport: people and goods flowed with the current down to Pittsburgh and beyond, and the city became a nexus for trade. Rivers, roads, railways, and canals flooded the Pittsburgh region with people and goods throughout the nineteenth century.[1]

New resources also fueled industrial development in Pittsburgh. The city developed at the center of a vast hinterland that served its needs for resources and markets. It was surrounded by the raw materials necessary for iron and steel production, including coking coal, the readily accessible bituminous coal of the Pittsburgh seam, and iron ore.[2] Perhaps more importantly, Pittsburgh sat at the center of markets for its industrial products, with population centers to the east, the Great Lakes transportation hubs to the north, and the expanding needs of the frontier and transportation opportunities of the Mississippi to the west. The confluence of markets, transportation networks, and readily available resources made Pittsburgh grow more quickly than its neighbors, and created an industrial giant out of the tiny settlement within a few decades.[3]

All of this development occurred despite Pittsburgh's less-than-ideal geography. The only flat land on which to build large industrial works was located on the point itself, or on the river meanders radiating from that point. Not far from the edges of the rivers, the land climbed precipitously into steep and rolling hills, making it difficult for rail or road to reach industrial sites unless they were located right on the water. One visitor wondered why Pittsburgh hadn't become the *most* populous and productive city of the nation, and eventually decided that "this it might have been, perhaps, if the site had been ten level square miles, instead of two, and those two surrounded by steep hills four hundred feet high, and by rivers a third of a mile wide. It is curiously hemmed in,—that small triangle of low land upon which the city was originally built." Pittsburgh's cramped geography came to magnify the negative impact of industrial production and domestic fuel consumption, both of which continued to grow in size and scope throughout the nineteenth century. The geography of the city ensured that any smoke produced by riverside industry would be trapped in the low-lying valleys or hollows. The effects of geography were exacerbated by meteorology on the occasions when cold, smoke-filled air at ground level was trapped in place by an overhanging layer of warmer air. Pittsburghers called the resulting inversions "Londoners."[4]

Pittsburgh's pall of smoke grew along with the city's population and industrial production throughout the early nineteenth century. Travelers and city residents wrote both to complain about the smoke and to excuse its necessity. As a result of the local domestic use of cheap and smoky bituminous coal from the Pittsburgh seam, travelers described Pittsburgh as a "smoky city" as early as 1800. Almost every visitor to the growing city remarked on the lingering smoke. One described it as "singularly gloomy."[5] The smoke increased in the second half of the nineteenth century as "the Pittsburgh region now emerged as America's center for crucible steel," according to historian Thomas J. Misa. By 1877 the region produced just under 70 percent of all U.S. steel. In short, Pittsburgh was quickly becoming a manufacturing center with national consequence. As *Harper's Weekly* enthused, "it is no exaggeration to say that nearly the whole world is laid under contribution to keep her immense and multifarious industries constantly supplied with the necessary material."[6]

The prodigious increase in industrial production was matched by an increase in the use of coal as primary fuel for almost all industrial and domestic applications, and by a slow degradation in the quality of available coal. This expanded need for coal drove more users to the "soft" bituminous coals, which produced greater quantities of smoke than the "hard" anthracite coals available in eastern Pennsylvania. Indeed, the combustion of coal was, at root, the cause of the city's notoriety. While other sources later contributed to the air pollution hanging over Pittsburgh (most noticeably the internal combustion engine), the massive amounts of coal burned in industrial and domestic applications were indisputably the source of the smoke that put Pittsburgh into a class by itself among polluted cities.[7]

The geography of industrial development and settlement intensified the effects of increased use of coal in Pittsburgh. Built along the confluence of the rivers, the city spread out in low-lying flats, irregular hills, and crowded hollows. Manufacturers chose mill sites to allow access to the rivers for transportation of raw materials, fuel, and finished goods; but that decision also insured that smoke-producing industry would be located in the lowest part of the valleys, where smoke could be trapped indefinitely. Smoke from stationary and mobile sources alike could lie for days in the valleys, trapped by the surrounding hills, choking the residents whose homes clustered in those lowlands. In Pittsburgh, coke ovens and company housing often existed side by side, and smoke and dust wafted over employees, their families, and the rest of the city's denizens. Compounding the effects of the problem, Pittsburgh continued to grow in population, from a recorded population of 49,221 in 1860 to 86,076 just ten years later, and an incredible 321,616 thirty years after that.[8]

The growing population, spectacular developments in industry, and increasing dependence on coal all contributed to the city's national reputation for smoke. One nineteenth-century account in *Harper's Weekly* became lyrical in describing the city scenes to a national audience: "The buildings up and down the river were black and dirty beyond all hope of cleanliness, the bridges were grimy with soot and smoke . . . and over all was a cloud of smoke that rolled and drifted and surged as thick as a black fog. Smoke was everywhere. . . . The streets, the business blocks, and even the private houses were unthinkably dirty and forbidding."[9]

Responding to Smoke and Pollution before 1969

The problems presented by smoke were obvious, but so were the challenges of controlling smoke. In particular, any attempts to limit smoke or pollution ran into the problems of government regulation of private property, the bewildering diversity of smoke sources, the unequal impact of the drifting pall, and an absence of a sufficiently strong government body to implement unpopular changes. As these difficulties and insufficiencies were confronted over more than a century, the fact remained that the general public had almost no direct influence upon attempts to regulate the air.

There were attempts to respond to dirty air during the nineteenth century, but they had little to do with the general public. Throughout the eighteenth and nineteenth centuries, courts allowed affected parties to bring suit against smoke as a nuisance, but nuisance law tended to prioritize productive industry. Nuisance suits were only a viable option for a very few members of the public. Generally only shop, factory, or farm owners who complained of economic loss due to smoke had standing to sue. Early nuisance law was entirely post hoc, that is, it depended upon the aggrieved party actually having experienced the nuisance before a complaint could be made; preemptive nuisance cases appeared later in the nineteenth century.[10] Sanitarian reformers of the mid-nineteenth century attempted to limit Pittsburgh's smoke as a health threat, leading to the first legislative attempts at smoke control in Pittsburgh. The city council passed an ordinance banning coke ovens within city limits and the use of soft coal in steam locomotives in 1869, but without providing for any clear method of enforcement. These were paper tigers, however: the city ordinance did not eliminate the several hundred ovens then in existence (or those built over the next hundred years) and the city council's ban on soft coal had no public support, little mechanism for enforcement, and almost no chance of success.[11]

It was not until the turn of the twentieth century that more meaningful regulation and the first involvement of the public in air pollution control matters emerged, as a part of what historians have come to call the Progressive movement. In Pittsburgh, the campaign to reform the city was coordinated by two Progressive Era organizations: the Women's Health Protective Association (or WHPA), founded in 1890; and its successor, the Civic Club of Allegheny County. These organizations, among others, were created by a confluence of interests reacting to the decay of urban centers. By the turn of the century, all types of pollution had come to mean more to city dwellers than just uncleanliness, or even a threat to health: urban pollution had come to be associated with moral decay, political corruption, disease, poverty, ignorance, and disorder. This perceived crisis spawned a reform mentality that brought together efficiency-minded engineers, civic boosters, and moral reformers, who together promoted efficient use of resources and the competent functioning of government in an effort to control urban disorder.[12]

In Pittsburgh and across the nation, Progressive movements featured the reform efforts of white, middle-class women as well as male medical and scientific experts. Women in Chicago, Pittsburgh, St. Louis, and Cleveland formed local organizations dedicated to improving the urban environs. These groups noted that women were often the first to experience the ill effects of air pollution. According to historian Carolyn Merchant, women had a unique role in reform efforts: "Nowhere has women's self-conscious role as protectors of the environment been better exemplified than during the progressive conservation crusade of the early twentieth century."[13]

Pittsburgh women moved to control smoke in the last decade of the nineteenth century. They were galvanized into action by a brief experience with clean air provided by a temporary local switch to natural gas, concerns over both the aesthetic and health effects of smoke, and the inspiration of the City Beautiful movement. Among other civic reforms, the WHPA actively lobbied for antismoke ordinances and strict regulation of railroad smoke while advocating for the expanded use of efficient, smokeless furnaces.[14] The WHPA argued for clean air based on its members' status as wives and mothers, but also used scientific evidence concerning the causes and effects of smoke. As one historian puts it, the "women turned to physicians and each other as experts on the smoke problem, for they knew no other groups in society who so well understood the effects of smoke on households and health."[15] The WHPA's white, middle-class women were mostly wives of prominent

Pittsburgh businessmen, but framed their argument in the terminology of morality and aesthetics as much as in the language of business efficiency.

The WHPA's successor, the Civic Club, implemented this basic argument with a more complex organizational structure. The club was a mixed-gender group, an important departure from the WHPA. A Civic Club organizer explained the reasoning behind the inclusion of men: "The men will provide the experience for the new club and the women the enthusiasm." Working in mixed-gender committees, the Civic Club focused its attention on a variety of municipal reforms, combining the expertise of its professional male members with the organizational strength of its women activists. Women and men held committee and club appointments jointly, and the club combined calls for the defense of the health of children with the professional evidence of male doctors and scientists. According to historian Loretta Lobes, the Civic Club described itself with self-consciously feminine characteristics: "The club's history identified one of the organization's chief characteristics as 'quiet persistency.' Furthermore the history portrayed the club as exhibiting a 'co-operative, rather than a critical, antagonistic spirit.'"[16]

The WHPA, the Civic Club, and later the Twentieth Century Club were instrumental in putting pressure on the city to enact and enforce city regulations for limited smoke control. Their Progressive organizing was the first example of successful public involvement in government policy on air pollution. But their efforts were largely limited to lobbying and exhortation; they were excluded from policy formation, implementation, and enforcement; and results were mostly restricted to voluntary action on the part of polluters. Progressive reform attempts were dwarfed as Pittsburgh was reaching the height of its industrial development. In the four decades following the 1890 census, the population of the city "nearly tripled," with the addition of thousands of new urban dwellers, part of a new wave of immigrants from southern and eastern Europe.[17]

At the same time, a new wave of expert reports, including the well-known Pittsburgh Survey and the Mellon Institute's Smoke Investigation, supported a drive for the passage of a variety of smoke regulations in the city of Pittsburgh. Following the ineffectual laws banning bituminous coal in locomotives in 1869 and eliminating coke production from the city limits, Pittsburgh's city council passed regulations with strict standards in 1895, which were nonetheless struck down by the courts in 1902.[18] A follow-up 1906 law prohibited "dense smoke" from industrial or commercial stacks. That law created a municipal Bureau of Smoke Control to investigate excessive smoke, levy fines, and recommend more efficient methods of burning fuel. Outside of the municipal government, a new blanket organization created in 1912—the Smoke and Dust Abatement League—brought together a variety of women's, businessmen's, engineering, and civic clubs under the rubric of the Chamber of Commerce to provide political support for the enforcement actions of the bureau.[19]

This prewar Progressivism had significant limits, however. The municipal bureau lacked significant enforcement power, and was often reduced to asking violators for voluntary compliance. Regardless of the efforts of civic associations, enforcement largely went by the wayside with increased production demands. "By World War I," says historian Martin Melosi, "when production of goods became a patriotic duty, smoke abatement fell on hard times." Pittsburgh thrived on the heavy industrial production necessitated by the war effort, and the largely educational and nonpunitive smoke control attempts were no match for patriotism in the war years, or fears of unemployment in the Great Depression. After decades of limited enforcement, the city's Bureau of Smoke Control was eliminated in 1939.[20]

It was not until 1941 that businessmen, civic boosters, newspaper editors, and city politicians banded together to adopt a new smoke control ordinance, following the lead of St. Louis. The combined force of these public figures was brought to bear on the previously intractable domestic smoke problem. Domestic use was staggering, with 141,788 households in 1940 burning coal in the city, and 53,388 of those using highly inefficient, manually fed coal-burning stoves.[21] But with the cooperation of powerful politicians and the support of the public, local newspaper editors were able to highlight the dirtiness of Pittsburgh's skies, the very real possibilities for improvement, and the political changes required for a successful smoke control campaign. Effective support from a few key public figures made for successful smoke regulation where the civic and voluntary groups "had never been effective," according to Joel Tarr.[22]

This time, however, support from the press, public health officials, and politicians made possible the creation of the Mayor's Commission for the Elimination of Smoke, which recommended the difficult but necessary steps that would have to be taken to curtail smoke production. The resulting ordinance, passed over the objections of Jones & Laughlin Steel and local coal companies, included observations of visible smoke using the "Ringelmann Chart" and requirements for domestic and commercial users to burn processed, "smokeless" coal, also known as "disco."[23]

The enforcement of this law was postponed for the duration of World War II, but by that point the city's air was shockingly dirty, and as historian Sherie Mershon points out, "the city's business and political leaders were ready to set aside old suspicions and rivalries to undertake a concerted program for urban revitalization." The concrete result of this postwar sense of public/private cooperation was the creation of the United Smoke Council, later affiliated with the politically powerful Allegheny Conference on Community Development (ACCD). Relying upon the persuasive powers of its civically prominent members, the council and the ACCD moved to regulate the burning of heavily polluting "soft coal" in industrial applications by 1946, and in domestic uses a year later. In its role as a "linchpin of a progrowth coalition," the ACCD brought together corporations and politicians who shared a vision of an economically burgeoning region, growing through and beyond the industrial roots of the local economy.[24] With the solid support of powerful businessmen and four-term Mayor David L. Lawrence, the city's new Bureau of Smoke Prevention embarked on an active career of management, inspection, and enforcement of the municipal ordinance. The Democratic machine—a system of reciprocal patronage and party-line voting—controlled much of the city payroll from the mid-1930s to the 1960s. Lawrence, first elected in 1945, quickly identified himself with a variety of urban renewal plans. This was known as the Pittsburgh Renaissance, an urban reform vision based upon an alliance between business elites and Democratic politics, complemented by an "interlocking directorate" of public and private technical experts and advisers. Because the Lawrence machine guaranteed political support for smoke control, it did not particularly matter if governing bodies did not include symbolic representatives of the citizenry; there was no political benefit to be had in including those individuals.[25]

Indeed, while the smoke control ordinances of the era did not specifically involve a great number of citizens in any role, city officials and businessmen did not require the support of citizens organizations or in-depth public involvement. Together, the ACCD and the city proceeded to enforce smoke regulation in areas previously thought impossible by reformers. Railroads, which had been exempted from county smoke control as "moving sources," were included in 1947 legislation passed with the support of an ACCD-forged "advance bipartisan consensus."[26] Allegheny County commissioners, empowered by the first piece of the Pittsburgh Renaissance legislation, created a seventeen-member advisory committee to write a new air pollution code. That committee was primarily stocked with industrialists rather than representatives of the general public. The members included representatives from the USC, labor unions, coal and steel industry, barge and railroad lines, other coal-burning industry, and "two women, one newspaperman, and one each from the Chamber of Commerce and the Pennsylvania Economy League," according to political scientist Charles O. Jones. Eleven months of negotiation created a consensus position and a county ordinance in 1949. This compromise was widely reported as an amicable consensus of both "industry and the general public" and as "the real beginning of voluntary action and support." Reform was so successful that the *smoke* problem seemed to have been solved (although the *pollution* problem had not yet been discovered). By 1950, at least according to Mayor Lawrence, Pittsburgh's air was clean, and everyone could thank the cooperation of the great men who sat down in a boardroom to craft a new, cooperative approach for smoke control. Members of the city's Bureau of Smoke Prevention met with power brokers, including Richard King Mellon and other corporate executives, all members of the ACCD. The resulting approach to smoke control emphasized the transition to low-emission coal in homes and factories as well as the elimination of residential burn barrels.[27]

By the mid-1950s the air over Pittsburgh was visibly cleaner, and Pittsburghers congratulated themselves. National and international observers also applauded the change and urged others to emulate Pittsburgh's example. The Smoky City had seemed to clean itself through regulation, enforcement, and fuel choice. Television network NBC broadcast a BBC-produced documentary entitled *Smokey Pittsburgh*, which declared victory over smoke. This film argued that for the first time in a century, Pittsburghers now could breathe clean air, thanks to remarkable cooperation between city government and area industry.[28]

The city's success prompted a 1951 change in state law allowing for the creation of county, rather than a state, health departments. This facilitated the creation in 1953 of a *county* agency, named, with a complete lack of originality, the Bureau of Smoke Control. By 1956 the city of Pittsburgh's Department of Health and its Bureau of Smoke Prevention were subsumed under county control when the County Board of Commissioners created the Allegheny County Health Department or ACHD. On January 1, 1957, the health department assumed responsibility for smoke control programs through its renamed Bureau of Air Pollution Control, commonly known as the Bureau. A new Citizens Advisory Committee was created to write comprehensive county smoke control laws. Once again the "citizens" that were a part of the committee were most likely to also be industrialists: "[T]his body—like its 1947 predecessor—was firmly dominated by corporate executives and technical experts," observe historians Sherie Mershon and Joel Tarr. In 1960 the committee drafted a smoke control regulation known as Article XIII, organizing enforcement centrally while covering a broader geographical area.[29]

While the county was ambitiously expanding its role, the Commonwealth of Pennsylvania moved much more cautiously, even in the wake of the nationally reported Donora incident. The state Department of Health finally wrote a very limited law in 1960 that attempted to balance regulation with economic development, and included yet another citizens advisory panel. After industry representatives objected to the proposed advisory health board that would have relied heavily on health professionals for counsel on regulating pollution, the state Department of Health created an Air Pollution Control Board that was instead "heavily representative of industry." The resulting eleven-member body relied on a "maximum of conference, conciliation and persuasion" in controlling pollution, and featured just one member of the general public not directly affiliated with industry or government. This was a very limited form of public involvement, however: that single member of the public showed up for just one of the first ten meetings before being replaced. Like the board it advised, the Technical Advisory Committee was stacked with industry representatives. The result was a lack of state action. During this period, the state abdicated the control of smoke to the cities directly affected, declining to intervene in any meaningful manner.[30]

On the national level, the first federal action to respond to public concerns about air quality was the Clean Air Act of 1955. This legislation focused on research about air pollution control under the oversight of the Public Health Service and its newly created National Air Pollution Control Administration. The 1955 act—the first to be named "Clean Air Act"—had no regulatory or enforcement component and did not intervene in state or local matters. The Clean Air Act of 1963, on the other hand, marked a permanent, substantive involvement in air pollution control by the federal government, one that actively encouraged citizen involvement in congressional hearings, local policy formation, and enforcement. As described in Chapter 1, Senator Edmund S. Muskie, chair of the new Senate Subcommittee on Air and Water Pollution, staged subcommittee hearings in cities across the nation, explicitly linking the federal program to local needs.[31] This unprecedented tour exposed the nation to the variety and complexity of air pollution issues in cities, and set the stage for an increase in citizen involvement in the following decade. Indeed, the 1967 Air Quality Act, which grew out of Muskie's tour, cemented a pattern: federal support for local action based on citizen involvement, rather than direct federal intervention.

Regulatory Success without the Public

If any representatives of the public wanted to become involved in pollution policy in Pittsburgh during the late 1960s, they no longer needed to work through the formerly dominant Democratic machine. By that time, countywide public sector institutions enjoyed greater power than the waning municipal government, and significant policymaking took place in the civic rather than the electoral arena in Pittsburgh. Thus, anyone seeking to influence policymaking would have to find a means to access institutions responsible for regional or countywide public planning. The county itself was quite large and politically complex: by the late 1960s the city of Pittsburgh was the largest of 129 separate municipalities scattered throughout Allegheny County.

This growth meant that neither the city nor the county could be reliably controlled by the urban machine, the tool previously wielded so effectively by Mayor Lawrence. When reformist Democrat Peter Flaherty beat the Democratic machine candidate in the 1969 mayoral race, he proceeded to restructure the city government and eliminate many of the patronage jobs in the city works department. "The net effect was to purge city government of many of its loyal machine troops," writes political scientist Barbara Ferman. Flaherty's actions strengthened the city's neighborhood groups at the expense of the business community, as "planning became more of a public as opposed to a private function." In dismantling the Democratic machine, Flaherty (and his successor, Richard Caligiuri) was part of a process that emphasized Allegheny County's government over Pittsburgh's.[32]

As the population of the city decreased sharply in the late twentieth century, patterns of suburbanization transferred political power from the city of Pittsburgh to the communities, boroughs, towns, and municipalities surrounding the central city. A growing bureaucracy administered Allegheny County, which covered over seven hundred square miles with a population in 1970 of 1.6 million. Every county in Pennsylvania had a similar system of government in the late 1960s, including a three-member, elected board of county commissioners. The commissioners controlled most of the county's services, including the county health department, the administrative home of air pollution control.[33]

Pittsburgh and its surrounding industrial centers reached their productive and demographic peaks in 1950, beginning a slow decline that lasted until the precipitous plunge of the late 1970s and early 1980s. Throughout that period, however, Pittsburgh was the corporate headquarters of a number of large international firms, including United States Steel, drugmaker Bayer Corporation, aluminum producer Alcoa, and PPG, the parent company of Pittsburgh Paints and Pittsburgh Plate Glass. Indeed, Ferman observes that "Pittsburgh was the third-largest headquarters center [in the United States], surpassed only by New York City and Chicago."[34] The ACCD took pains to highlight the transformation of the city from industrial workshop to corporate headquarters. The ACCD specialized in public-private partnerships to address the smoke problem, plan urban housing, and rebuild Pittsburgh's downtown. The combined efforts of the ACCD and regional leaders were widely praised throughout the 1950s and 1960s as a laudable example of successful, collaborative planning.[35]

The ACCD's de facto leadership of regional planning did not go unchallenged, however. By the end of the 1960s Pittsburgh citizens condemned the public-private partnership that marked the ACCD's previous attempts to clear the air as improper collusion between regulators and industry to mitigate the impact of pollution control. Many citizens publicly demanded the exclusion of corporate interests from pollution policy or at least the counterbalancing involvement of citizens in a variety of city planning and policy issues. By 1969, according to historian Sherie Mershon, "Conflicts between political and business leaders and among private civic associations undermined elite consensus. . . . Groups that had previously stood outside the governing coalition, most notably African Americans and neighborhood-based community organizations, forcefully asserted that the ACCD-centered progrowth coalition had ignored or harmed their interests."[36]

The debate spilled out in heated public meetings, editorial screeds, letter-writing campaigns, and publicity stunts. Unprecedented pressures from federal law and local activism dumbfounded corporate leaders, who were accustomed to years of successful cooperation with local government. Pittsburgh-based U.S. Steel "and the other steel companies grew increasingly irritated because they were being pushed over the cliff of diminishing returns," says historian Brian Apelt, presenting the industrialists as beleaguered by outside forces: "Still, environmentalists kept raising the bar."[37]

The issue of air pollution control escalated through the hierarchy of government in the 1970s. What had previously been a matter of municipal law became the province of counties, states, and even the federal government. The location of air pollution law was a negotiated compromise. When any party did not benefit from the debate on the city, county, state, or federal level, it could seek to change the venue. In Pittsburgh, for example, citizen activists and city politicians sought to move air pollution regulation to the county level to better manage a large airshed and to gain regulatory power. Corporations that did not benefit from county regulation sought to move the locus of debate to the municipal level by appealing to local politicians on the basis of job creation, to the federal level to seek more uniform or moderate regulation, or to the federal courts to overturn county decision making.

The definition of the pollution problem also escalated in the postwar period. While Pittsburgh's leaders and journalists had declared victory in the battle to control *smoke* in the 1950s, the battle to control *air pollution* was not only incomplete but also largely undefined until the 1960s. City government and the ACCD had focused on visible smoke and its aesthetic impacts in their campaign against smoke, and rarely mentioned invisible pollutants. This was either a result of an inherent bias toward mitigating the impact of control or of limited knowledge concerning the specific public health impacts of air pollution.[38] But a series of widely publicized events brought attention to unseen threats to public health. From the fatal and widely reported air pollution disaster of Donora, Pennsylvania, in 1948, to revelations concerning radioactive fallout from nuclear testing in the late 1950s, to Rachel Carson's 1962 publication of her bestselling *Silent Spring*, all emphasized the unknown complexities and unintended consequences of chemicals or contaminants in the atmosphere. In Pittsburgh, the attention of regulators turned from visible dust and smoke to the invisible impact of carbon monoxide, sulfur, and nitrogen oxides and trace contaminants detectable only by the use of monitoring equipment. Over time, the health impacts of these and other pollutants became a primary concern of activists and regulators alike.[39]

State law in the 1950s and 1960s located the power to enforce air pollution regulations in the county health departments. Under the Local Health Administration Law, Allegheny became the only county in Pennsylvania to take advantage of this power, creating its own air pollution enforcement organization independent of the state. The county health department thus replaced the city's board of health in the active enforcement of Article XIII (1960), described as "the basic legal machinery" for control of air pollution in the county. Article XIII was a part of the Allegheny County Health Department Rules and Regulations passed under the controlling state legislation. Pollution sources in the Pittsburgh region came to be regulated by a three-level hierarchy of air pollution law: county, state, and federal agencies each had jurisdiction over the activities of polluters in the city of Pittsburgh.[40]

At the time of Article XIII's passage in 1960, county commissioner William D. McClellan boasted that "the regulations would be 'the most stringent governing any industrial area in the United States.'" Indeed, most journalistic accounts used the same word to describe the law: "'Toughest' Smoke Law Promised," wrote one, "County OKs 'Toughest' Smoke Bill," another, and "Toughest Smoke Law Voted," a third. The passage and enforcement of this "tough" law was only possible with the political support of the ACCD, which continued to emphasize voluntary controls and cooperative regulation.[41]

Though commentators in the early 1960s thought the new rules to be "tough," what Article XIII lacked for the action-minded environmentalists of the late 1960s was any serious provision for public involvement in the pollution control process. This desire for greater public involvement was evident even as Article XIII became law, though to no avail. Mrs. Thomas Horrocks, the secretary of a short-lived citizens group named the Allegheny County Citizens Against Air Pollution (ACCAAP), wrote to Dr. John D. Lauer, chair of the county board of health (at the address of his employer, Jones & Laughlin Steel), in complaint: "Under the terms of this ordinance, the Advisory Committee is required to report only to the Board of Health; there is no provision for reports being made to the County Commissioners or to the public. . . . Toward this end, we again recommend that a citizen representing the general public be given a place on the Air Pollution Advisory Committee."[42]

The pleas of the ACCAAP were not entirely ignored in the creation of Article XIII. County commissioners leavened the membership of the advisory committee with the addition of a single citizen not representing a professional or industrial organization, something the press found noteworthy. "Housewife Gets Post," said the *Press* headline, and the article observed that "the biggest surprise was the selection of a housewife, Mrs. Jean Nickeson." The county commissioners and journalists alike highlighted Mrs. Nickeson's role as a housewife. Under the heading "Housewives Work," the *Press* highlighted Nickeson's perspective: "Pointing out that 'It's the housewives in Allegheny County who must clean up the dirt which pours from industrial smokestacks,' [Nickeson] urged that the homemaker be given a voice on the committee which sets up compliance schedules for local mills. 'Speaking for myself,' she said, 'I'm a little tired of serving as an unpaid clean-up woman for industry.'"[43]

While the addition of a housewife might have been deemed noteworthy, Sherie Mershon argues correctly that "the citizen group [the ACCAAP] had no substantive impact on county policy," and the committee they advised was not that significant itself. The members of the advisory committee, housewives or otherwise, were not specifically involved in rule making or enforcement. The advisory committee met throughout the decade, but without a specific charge or delegated authority from Article XIII, the committee's activities were limited to exhortations for voluntary action on the part of citizens and corporations. Their actions were surely heartfelt and well meant, but they were also the opposite of substantive involvement. In general, the committee's actions mostly centered on voluntary beautification efforts. They reached out to the community with printed place mats distributed to local restaurants and filled with facts about county air pollution control efforts. They also attempted to inform the public through essay contests, bus tours, and a yearly "Cleaner Air Week" publicity effort, often culminating in the release of bunches of yellow-and-black balloons at the Pittsburgh Steelers football team halftime shows. These publicity efforts were coordinated by the wives of committee members, funded by the ACCD, and largely attended by the wives of business leaders.[44] The minutes from a 1965 meeting of the advisory committee captures the tone of these events: "Mrs. [Jean] Nickeson reported on the Cleaner Air Week program for 1965. She informed the Committee that she held the first annual Cleaner Air Week tea at her home for members of her community. . . . Chairman Weaver suggested that wives of the members of the Advisory Committee might consider similar teas for Cleaner Air Week in future years."[45] No matter how powerless the advisory committee was, the local press reported favorably on its actions, especially during the annual "Clean Air Week."[46]

National observers likewise praised this very limited inclusion of the public in Pittsburgh's air pollution control mechanisms. In general, public health professionals expressed support for at least the nominal inclusion of the public in air pollution matters. But some went much further. As noted earlier, an article in the *American Journal of Public Health* specifically offered the example of Pittsburgh as an air pollution success story, and theorized that the inclusion of the public in air pollution policy matters satisfied two core American political beliefs: "From the idealized Greek polis, to the idealized New England town meeting, to the idealization of citizen participation in modern metropolitan government affairs, the lineage and vitality of these two values—decentralization of political control, and pluralism of political initiative—show themselves very clearly."[47]

An overreaching pamphlet from industry lobbyists the National Association of Manufacturers declared the limited inclusion of the public a sweeping success. For the NAM, presumably, any example of cooperative action provided an important bulwark against unwanted federal intervention: "No law at any other level—national or state—could have possibly evoked the enthusiastic response and compliance that the Allegheny County ordinance was accorded," the author wrote hyperbolically. "Industry and general public readily identified with the local ordinance because it was largely their creation; not one dictated from afar by legislators totally unaware of the true nature of the problems peculiar to the community."[48]

For the NAM the issue was not so much public involvement in policymaking, but local control of policymaking—not necessarily the same thing. Even while praising voluntary air pollution control and public input, the pamphlet lists the members of Allegheny County's advisory committee, which consisted of industry representatives and public health experts. The chairman was from Westinghouse Electric Company and the remaining members were from Duquesne Light, U.S. Steel, the Pennsylvania Economy League, Consolidation Coal, the Pennsylvania Railroad and Union Barge Lines, industrial research centers at the Mellon and Carnegie Institutes, the University of Pittsburgh's School for Public Health and School of Medicine, and the United Mine Workers. Experts and industrialists there were aplenty; but where was the public?

Despite the minimal citizen representation and the occasional complaint from the Allegheny County Citizens Against Air Pollution, there seemed to be little reason to protest the substantive exclusion of the public in air pollution policy matters, as most journalistic coverage from the early and mid-1960s was especially supportive of Pittsburgh's control efforts. As in the case of the NAM publication, observers specifically praised public input as a central part of policy formation and implementation. Bureau chief Edward L. Stockton addressed the 1966 Pittsburgh meeting of the American Industrial Hygiene Association to argue that while the battle was not over, "Allegheny County citizens are prepared . . . to continue their anti-pollution campaign with unrelenting vigor, secure in the knowledge that they can depend on the same measure of cooperative community action that brought them this far to carry them the rest of the way."[49] Stockton could not foresee the upheaval and controversy that transformed air pollution control. The era of cooperation soon came to an end.

Getting in the Door: GASP and Air Pollution Control after 1969

Until 1969, then, public involvement in air pollution issues was limited in scope and effect. While many observers commented on smoke, and some complained, citizens themselves were almost never admitted to policymaking bodies or placed in politically powerful positions. From a historical perspective, citizen involvement in pollution control was limited to those elites positioned to comment on smoke's effects in published accounts or, later, through the lobbying efforts of civic associations. Seen in this historical context, the development and substantive inclusion of citizens organizations in the regulatory process after 1969 is a remarkable and significant departure from precedent. Activists, politicians, and administrators alike were empowered by the 1960s movement for citizen involvement in public affairs, and they combined to fundamentally transform the importance of public participation in the formation of environmental policy and the mechanism of air pollution control. Before that could happen, however, activists would have to "get in the door"—they would have to gain real, substantive access to policymaking and decision-making bodies.

In Pittsburgh as well as throughout the United States, the rise of modern environmental concern resulted in a battle to reshape the interaction of citizen and state. Conflict over air pollution control fed upon itself: environmental activists sought to educate and inform, mobilizing more citizens, who in turn sought further and more substantive participation in policy decisions regarding air pollution control. More involvement by the citizenry brought more conflict, more media attention, and more members, starting the process anew. As historian Shawn Bernstein has noted, "Escalation of conflict further intensified public and media interest and involvement. More and more groups were mobilized with a stake in air pollution control as an issue of policy."[50] In comparison to the exclusive, elite-driven attempts at smoke control from the first half of the twentieth century, air pollution control after 1969 was a tumultuous, but nonetheless inclusive, affair.

Citizen involvement in air pollution policymaking and day-to-day enforcement in Pittsburgh was mostly the result of actions taken by GASP—the Group Against Smog and Pollution—the local environmental organization formed in 1969 as a direct result of federal regulatory measures. GASP's creation and fight for inclusion is highly illustrative of a transition in air pollution control, from a period characterized by the involvement of elite groups and corporate self-regulation to a hotly contested public battle involving media coverage, protest, and legal pressure. The distinction here is between a period when citizens were only a symbolic part of the public air pollution debate and a period when citizens were a part of public debate and were also involved in "substantive" actions: reviewing, criticizing, opposing, promoting, or bringing suit against both regulators and regulated.

For GASP to be included in the institutions of air pollution control, however, public participation had to be legally redefined. *Direct interest,* a legal term limiting participation to those who could demonstrate an economic involvement in the matter before a court, was expanded to include the public interest, a change that superseded the previous categorization of pollution complaints as a part of the law of nuisance. This redefinition of public involvement in Pittsburgh preceded a national transformation of the importance of public involvement in policy matters, and prepared the way for a further change in the legal standing of citizen groups.

In 1969, the commonwealth of Pennsylvania and Allegheny County each conducted public hearings on proposed revisions to the state air pollution law and the county rules and regulations. The federal Air Quality Act of 1967, which allowed states to set their own air quality standards within the federal framework, required public hearings as a part of the formation of new pollution laws. In fact, the act required the state to "adopt, after public hearings, ambient air quality standards" applicable to the state's air quality control regions. It was up to the state to propose, evaluate, and create the air pollution control regions and the air quality standards in order to receive federal support for air pollution control.[51] Thus, federal law required citizen participation in public hearings on local air pollution control matters, and the National Air Pollution Control Administration (an agency under the direction of the Department of Health, Education and Welfare) distributed materials encouraging the formation of local organizations that could speak knowledgably on air pollution issues. In the NAPCA's words, "The Clean Air Act provides opportunities for all who are concerned with air pollution to participate in the decisions being made about the quality of the air we must all breathe. These decisions will more often reflect community needs and values if the public will take advantage of these opportunities."[52]

The NAPCA granted funding to the Pittsburgh chapter of the League of Women Voters to organize and prepare interested parties to appear at the hearings, held in September and October 1969. The League, the Western Pennsylvania Conservancy (a conservation-minded local group founded in 1922), and a local chapter of the Federation of American Scientists organized a series of seminars and encouraged those who came to speak at the meeting organized by the Allegheny County Air Pollution Advisory Committee described at the beginning of Chapter 1.[53] Together, these groups worked to prepare speakers who could cogently argue for strict regulation. The result was a shock to the normally sparsely attended, industry-heavy public hearing process. Organizers had to scramble for more space, and the flood of speakers forced extra hearings that lasted into the night. As recounted in Chapter 1, local newspapers reported that "the roof fell in" on the committee as hundreds of citizens appeared to excoriate the proposal as "legalized murder."[54]

At the same time, other hearings at the state level had also been the site of a surprising citizen outcry, according to the Western Pennsylvania Conservancy: "An unprecedented number of people appeared at the hearing, speaking individually or as representatives of civic and conservation groups, medical societies, unions and other groups. Vocal citizens nearly unanimously attacked the proposals as too weak and suggested more stringent standards be adopted."[55] Transcripts of these state hearings register the surprise of the state officials. Air Pollution Commission chair Victor Sussman had to scramble to find a public space large enough for the Pittsburgh meeting: "It appeared late yesterday a great number of people were going to appear here, and [Joshua] Wetzel [of the Conservancy] and his staff helped us by very graciously helping us to get this meeting room," he noted at the outset. The "great number of people" who spoke lacked only torches and pitchforks to make them a stereotypically angry mob: "The standards for Pennsylvania that are suggested are little worse than a license to kill. They must not be tolerated," said Dr. Richard Lund. "I am sick and tired of glib industrial excuses and polished talks prepared by smooth talking public relations specialists," argued Stephen L. Gomes, a Ph.D. student at the University of Pittsburgh. "I'm sick and tired of the so-called relational approach. The people need an action approach. The time for talk is over." The transcript lists a Mrs. J. Lewis Scott, speaking for the "Carnegie Institute research center," who declared that "the cesspool in the sky must not be permitted any longer."[56]

After testifying at the Allegheny County hearings, some Pittsburghers proposed a more permanent organization of concerned citizens, with the eventual aim of providing ongoing oversight for similar hearings. A group of forty-three put their names on a list at the hearings with the intent of forming an air pollution watchdog group. Participants at an organizational meeting at the Squirrel Hill home of Michelle Madoff chose the name Group Against Smog and Pollution, or GASP. Those early organizers distributed a letter seeking a wider membership and addressed to all "Fellow Breathers." This letter described a diverse membership: "GASP is composed of citizens, teachers, scientists, physicians, housewives, businessmen, union leaders, senior and junior citizens." The letter listed the goals of the organization: "to testify and press to [*sic*] the same regulations as Saint [*sic*] Louis and other cities that are adopting to protect their people, and the generation yet unborn." Though the letter laid claim to a diverse membership, it listed three white, middle-class women as contacts: Madoff (listed as chair), Arlene Nadelhoff, and Nancy Bowdler.[57]

GASP's claims to represent a large number of residents, combined with the surprisingly large turnout at the public meetings, provided support for the passage of what were widely described as the nation's most stringent air pollution laws.[58] The *Wall Street Journal* described GASP as a forerunner of a new activist movement: "a curious collection of unionists, conservationists, health societies, ladies' garden clubs and college-age militants—the so-called breather's lobby." The *Journal* may have been taking a derisive tone in associating the League of Women's Voters with "ladies' garden clubs," but it was, nonetheless, correctly identifying a component of the new environmental movement, most certainly a "curious collection" of previously unheard-from groups and organizations. By joking about the existence of a "breather's lobby"—as if something so universal and unremarkable as breathing could undertake its own interest-group lobbying alongside long-established political interests—the *Journal* unknowingly had hit the nail on the head.[59]

The new county air pollution code established after the fall 1969 hearings, known as Article XVII, provided an important role for citizens and community action. Article XVII guaranteed the public a greater voice on the advisory committee, a limited right to speak on matters before the board charged with implementing the code, and the ability to receive more information from the ACHD and its divisions. The difference in philosophy between the new article and those preceding it is dramatically obvious from the text of the regulations. The 1949 regulation stresses cooperation above all else, and specifically identifies industry as a stakeholder in policymaking without mentioning the general public: "[T]hrough research and cooperative effort, the Director and the Committee shall continually strive, in collaboration with representatives of all affected groups and industries, to diminish air pollution and promote sound air pollution control practices."[60] *Collaboration* in *diminishing* and *promoting* is a fairly conciliatory approach to a limited goal.

Nothing much changed in this general statement of policy in the 1960 revision. But the text of the 1970 regulation is quite different, emphasizing the responsibility of industry, prioritizing the health of the public, and making a strong statement concerning "all available methods":

> *It shall be the policy of the County of Allegheny, in cooperation with Federal and State authorities, industry and other interested groups, to achieve and maintain purity of its air resources consistent with the health, welfare, and comfort of the residents of the County and the protection of property and resources, and to that end to require the use of all available methods of preventing and controlling air pollution in the County.*[61]

In one short decade, both the goals and the methodology for the project of air pollution control had changed. These two developments were not unrelated; with the significant inclusion of the public in air pollution control, the days of top-down reform were over. Future air pollution control policy would be substantially different because it would be collaboratively formulated and enforced by representatives of industry, government, and for the first time, the public.

As mentioned above, the politically contentious mechanism of air pollution control was removed from city government as the city was undergoing a transition from decades-long control by the Democratic machine of David L. Lawrence. "Davey" Lawrence served from 1945 to 1959, but his hand-picked replacement, Joseph Barr, continued his policies until 1968. Democrat Peter Flaherty beat Lawrence's machine candidate for mayor in 1969, and "Adopting the slogan 'Nobody's Boy,' . . . declared his independence from the machine, charging it with waste and inefficiency," according to political scientist Barbara Ferman. In turning his back on the machine, she writes, "Flaherty endeared himself to the neighborhood movement, which had become highly critical of the city's famed 'public-private partnership.'" With the machine deposed, there could be no continuation of the previous mayor's cooperative relationship with business elites: Flaherty "gave notice that the days of 'business as usual' were over" by restaffing the city government and its urban planning arm with technocrats who lacked the previous administration's close ties with the ACCD and business communities. The diminution of the ACCD's power in the mayor's office was echoed in the redesigned process of air pollution control: the cooperative, public-private partnership had fallen from favor, to be replaced by pluralistic competition between citizens, industry, and government.[62]

The new Article XVII provided an important role for citizens and community action, replacing the role taken by the ACCD and industry representatives in previous smoke control efforts. Article XVII established the Health Department's Board of Air Pollution Appeals and Variance Review, informally known as the "Variance Board." Companies or organizations that were found to be polluting in violation of county code could appeal for a variance or face a $100-per-day fine. The five-member board could subpoena evidence, hear appeals and testimony from the violator or the Bureau, and hold public hearings. GASP seized the opportunity for involvement and later boasted that not only had the group recommended four of the five members of the original board but that it had also lobbied to establish an appropriate pay scale that would help the Variance Board attract qualified applicants.[63] At the same time, GASP (in the person of Michelle Madoff) won a permanent seat on the Allegheny County Air Pollution Control Advisory Committee, which advised the ACHD Bureau of Air Pollution Control. While these accomplishments were important, they were nowhere near as significant as what happened when GASP attempted to go beyond the rights that were explicitly granted in Article XVII.

Citizen Standing, GASP, and the Variance Board in 1970

Through the early 1970s GASP organized itself into a dynamic, broad-based grassroots environmental group. GASP capitalized on the social networks of middle-class women and professional expertise of its membership to maintain links with churches, educational institutions, community groups, and labor unions. By the mid-1970s, GASP claimed a ready pool of professional talent and a membership of forty thousand from affiliated organizations. In its opening years, GASP members were particularly interested in sulfur dioxide levels in ambient air, which focused the group on industrial air pollution rather than automobiles. Additionally, GASP promoted public knowledge of air pollution levels and health risks, and used that public interest to argue for strict enforcement of existing laws.[64]

GASP was particularly interested in the Variance Board. This semiautonomous hearing board, created as a part of Article XVII, was designed to ease the transition to the new county code and was arguably the most visible component of the new regulatory regime. Michelle Madoff goes so far as to claim credit for the creation of the Variance Board, saying that she approached the county commissioners: "I went to the boys and I said, 'Fellas, you got to do something about air pollution, and you don't want to do anything, but you don't want to shut down any industry, right?' and they said 'right' and I said 'Why don't we create a committee that represents the best and the brightest . . . and they make the decisions and you got clean hands!' They loved it."[65] Once established, the five-member board was prepared to hear testimony from the public before making decisions about granting polluters variance from county law, and to allow any "interested party" to take a more active role in cross-examining witnesses, submitting technical reports, and making motions. GASP was certainly interested in this more active role, but the corporations whose variance petitions came before the board protested against its more substantial involvement. As the first GASP newsletter noted, "[a]t almost all hearings the attorneys for the major polluters go on record as objecting to GASP's role in the hearings."[66]

Duquesne Light, a local utility seeking a variance at the board's first meeting, objected to GASP's presence in general, "and to Michelle Madoff in particular," according to Madoff.[67] The utility most likely approached the Variance Board and GASP's involvement with some trepidation in the summer of 1970, as it had already had previous run-ins with the group and the press it could generate. "Power Firm Chided on Air Problem" read a *Post-Gazette* headline from January of that year, as GASP experts showed up at a hearing of the Public Utilities Commission to argue against Duquesne Light's request for a rate hike. "If you were to clean up your share of the filth we breathe . . . more customers would flock to the area," Madoff lectured the utility. The three-day long hearing had grown contentious, as Duquesne Light lawyers objected to being cross-examined by intervening parties, declaring that they were neither criminals nor negligent. "Apparently you are," replied an attorney for the U.S. Atomic Energy Commission.[68]

It therefore was to be expected that Duquesne Light would have doubts about the Variance Board, its practices, and GASP's participation. To begin with, company officials were concerned that being required to give testimony about the utility's emission levels would violate their constitutional right to avoid self-incrimination, an objection that Variance Board chair Robert Broughton brushed aside: "I must admit I remain somewhat mystified as to your legal justification," he wrote in response to the utility's objection. "A number of cases have held that a corporation cannot claim the protection of the 5th Amendment." Still, since he observed that "GASP intends to press this issue," Broughton asked for briefs on the matter, prompting the utility to withdraw the objection.[69] Lawyers for the utility also wrote to Broughton to complain about the news that "various activist groups are planning to make mass appearances" at the upcoming Variance Board hearing, and that this would be detrimental to "proper and orderly decorum." After suggesting changing the venue in order to better manage the expected crowds, attorney David M. Olds went on:

> *Please be advised also that we object to the intervention of the organization known as Group Against Smog and Pollution as well as the Federation of American Scientists. We would expect that each of these groups in view of our objection would have to prove their direct interest in these proceedings at the time of the July 20 hearing as a first order of business. Under the Local Agency Law, we submit that unless there is such direct interest, there is no right to intervene as a party.*[70]

In the narrowly defined sense of "direct interest," Duquesne Light's attorney was correct: Article XVII declared that "any person may appear and testify at a hearing," but that a more substantive status was reserved for two types of parties: "At an Appeals Board hearing the parties involved and the Director may appear with counsel, file written arguments, offer testimony, cross examine witnesses, or take any combination of such actions."[71] So, while any person had the right to speak at a hearing, only the Bureau director and "the parties involved" had a more direct role. As Olds pointed out at a Variance Board meeting, "parties involved" appeared to mean those with standing, or a "direct interest"—and that, he argued, had been interpreted by Pennsylvania courts "to mean an economic interest—as distinguished, for example, from a political or social interest."[72] The new Article XVII allowed parties without an economic interest to appeal decisions from the Variance Board, but not to present evidence, testimony, and reports to sway the board decision, and certainly not to cross-examine industry witnesses. But GASP, along with a few other groups and government entities, wanted to take this more active role. GASP challenged its potential exclusion in two ways: by pointing out that the new county ordinance required public input from anyone affected or "bound" by board decision, and by disputing the validity of a narrow definition of direct, economic interest.

Art Gorr, GASP's pro bono lawyer, wrote to Broughton in response to both Duquesne Light's concerns about the size of the expected crowds and the issue of direct interest. The two issues are ironically intertwined in his response, highlighting the fact that Duquesne Light was objecting to public participation in the face of large and threatening crowds. In a strange twist for the group, Gorr promised that GASP "has been working with interested citizen groups to hold down the numbers in attendance." Portraying GASP as barely holding back the enraged hordes, Gorr points out that in a previous board hearing . . .

> *we were able to limit the attendance of irate citizens to about two hundred, and very few people had to stand. Fortunately the hearing lasted until nearly 1:00 a.m. and those waiting in the hall were finally able to sit in. . . . Furthermore we were able to pacify them when they learned that only a few of them could testify in person. While we cannot guarantee the same success in the case of Duquesne Light, we will do our best.*

By invoking the need to pacify the irate citizenry into the wee hours of the morning, Gorr adds completely unsubtle weight to GASP's argument for its consideration as a direct interest: "We are all directly affected by the dirt and SO_2 which Duquesne Light is dumping into the air and into our lungs. We are bound by the action of the Board. We feel our intervention is in the public interest and we believe that the County Board of Health would agree."[73] Speaking to the *Press*, Gorr was a bit more blunt: "It is a very cynical view of the law to argue that any direct interest is a financial interest," he said. "If the people who breathe the polluted air coming from Duquesne's stacks don't have a direct interest I don't know what a direct interest is."[74]

Broughton ruled that GASP, the commonwealth of Pennsylvania, the city of Pittsburgh, and the Federation of American Scientists each had an interest in the proceedings and could thus testify before the board. In doing so, he overturned Pennsylvania precedent. Duquesne Light's attorney certainly thought that Broughton's decision was a significant break with the past. As David Olds noted in a letter admonishing the editors of the *Press* for perceived failings in their coverage of the affair, "the chairman of the board, in overruling the objections of Duquesne Light, stated publicly that he was adopting a new definition of 'direct interest' not to be found in any decision of the Supreme Court of Pennsylvania." Olds demanded an apology from the *Press* for their coverage, which he felt unfairly sided with the public in ignorance of the law; the *Press* was entirely unapologetic.[75] But the precedent was set. In a report to the advisory committee Broughton cemented Duquesne Light's defeat by stating a particularly expansive interpretation of his decision: "Any person or organization may petition the board to intervene and appear and testify at a hearing."[76] The phrase "to intervene" is important; Broughton was not simply allowing the public to speak but to act as a party with direct interest and therefore the ability to cross-examine witnesses, submit evidence and briefs, and make motions. Broughton built on this precedent himself, issuing a similar ruling in a different context several years later. After being appointed chair of the commonwealth of Pennsylvania's Environmental Hearing Board in 1972, Broughton ruled that "Environment Pittsburgh, a citizens group that petitioned to intervene" in a state Department of Environmental Resources suit regarding U.S. Steel, had a "right to participate in all aspects of the case . . . 'as if it were a party.'"[77]

Even before Broughton's decision, the mere fact that Duquesne Light challenged public involvement meant that it lost ground in the court of public opinion. A July 27, 1970, editorial in the *Press* had sparked a public debate when it noted that

> *Duquesne Light, a major polluter which seeks exemptions from new, strict smoke-control regulations, contended the public had no legal right to take part in the hearing since it lacked direct interest. . . . The Electric firm was the first to challenge public participation in appeals proceedings. Its argument that dollars take precedent [sic] over health suggested the old 'public be damned' attitude.*[78]

GASP members flooded the local press with responses. A letter from Madoff asked if Duquesne Light felt threatened by GASP:

> *Could it be because, at previous hearings, GASP attorneys have asked the right questions, and separated fact from fiction–which certainly played a role in the tough decisions . . . ? Could it be because, for the Duquesne Light hearings, GASP has gathered the talents of specialists in heat transfer, metallurgy, analytical chemistry, electrical engineering, economics and environmental control?*

A telegram from GASP member Pauline Nixon similarly supported both GASP and the "adversary format [which is] simply keeping the variance hearings balanced and more honest." A single supporter of Duquesne Light's point of view wrote tersely that the *Press* "confirmed my prior impression that most reporters are ignoramuses."[79] In the alternative weekly the *Pittsburgh Forum*, a GASP member also thought that Duquesne Light was attempting to silence GASP for asking troublesome questions, arguing that Arthur Gorr's "questions, and those of the city attorney and state officials, incited [Olds] to request that GASP and the others be barred from the hearings." The author was delighted that "Olds went down for the count."[80]

The opportunity to act as an "interested party" gave GASP the ability to cross-examine, challenge, and criticize industrial representatives before the Variance Board—not in separate testimony, but directly questioning representatives of government and industry before the attentive members of the board, with answers transcribed in the official record. In this role as an active intervenor, GASP representatives directly challenged those seeking variance from air pollution law, demanding specific answers about technical and economic decisions within the offending company. As the utility had feared, GASP particularly focused this power on Duquesne Light: "Foes Fired Up to Smoke Out Power Firm," read a story previewing a coming Variance Board hearing in November of 1970. "A major confrontation is expected tomorrow," went the story, as "on hand to oppose the requests will be attorneys for [GASP] and several other conservation organizations." The key point here is that the public would be showing up not merely with placards and protests but armed with that most fearsome of weapons, the lawyer. As the fight raged on, the utility was increasingly singled out for its contribution to pollution: "Lung Peril Charged to Duquesne Light," blared a *Post-Gazette* headline later in November, and "Conservation Group Rips Duquesne Light Efforts," read a *Press* story, as GASP experts and others declared that more than a quarter of the total sulfur dioxide in the county's air came from Duquesne Light's coal-burning facilities.[81]

This active intervention by representatives of the public is quite remarkable, and cannot have been warmly welcomed by those on the receiving end of this treatment. Some even doubted the group's motives. A former deputy director of the ACHD later argued that GASP was dedicated, but "dedicated for maybe their own ends. Michelle Madoff is a beautiful case in point. You know, she parlayed her membership in GASP into a seat on the city council."[82] Despite the resulting friction, GASP pioneered a novel legal right for the public, a new definition of legal "standing" that could make further citizen involvement in air pollution regulation a possibility. Robert Broughton's decision to go beyond what was specifically allowed in county law to allow substantive intervention in the Duquesne Light hearing was the first real test of a broader definition of inclusion that exceeded what was outlined in Article XVII. This was a meaningful, structural transformation of policymaking in response to the impassioned outpouring of public outrage in the fall of 1969.

This redefinition of citizen standing was arguably even more expansive in the appeals process that could follow Variance Board decisions. Article XVII noted that appeals to Variance Board decisions could "be made by any person suffering legal wrong or adversely affected or aggrieved by the decision." This was a much broader definition of standing to appeal than previously defined by the Pennsylvania Administrative Agency Law, which stated only that "Any person aggrieved thereby who has a direct interest in such adjudication shall have the right to appeal therefrom." Again, the courts had previously interpreted "direct interest" as requiring a financial or economic connection or harm stemming from the case. But now, as local attorney Robert S. Bailey noted in a 1970 article in the *Duquesne Law Review*, "the Allegheny County regulations . . . [seem] to confer standing to appeal on those who may not be so directly affected, but who still may have a valid complaint." Bailey argued that "persons aggrieved" had been defined in previous cases as meaning those who owned property in the county (those who held a direct financial interest in the continued value of the property, in other words) while excluding renters, tenants, or commuters. "In this light," writes Bailey,

> *the County Regulation's use of "person suffering legal wrong" as a criterion for standing also appears to imply further that there be no need for land ownership. Thus, by its own language, it appears that residence or possibly mere occupational location within Allegheny County, coupled with a showing that the variance grant will affect the quality of the air the person breathes, could be all that is required to appeal to Allegheny Common Pleas Court regarding the grant of a variance.*[83]

Robert Broughton's decision to allow GASP to act as an "interested party" before the Variance Board set a precedent in Pennsylvania, and possibly the nation. Broughton's decision may be significant as the first administrative decision in the nation to allow citizen standing not explicitly based on a citizen suit provision. It was, instead, based upon the *rejection* of an economic definition of direct interest. Although 1965's *Scenic Hudson* Supreme Court case established a legal right of intervention outside of direct interest, that ruling was based upon the provisions of the regulations governing the Federal Power Commission, which specifically allowed for public involvement. In making his decision, Broughton did not rely upon legislation that specifically allowed public input. Instead, he contravened the state regulations that should have governed the issue of standing, in itself a clear break with precedent. Buoyed by the developing philosophy of citizen standing and a rising emphasis on the importance of public participation, Broughton's decision is an important but forgotten part of the slow rise of citizen suits, which became steadily more important as a means of enforcing environmental legislation throughout the

1980s and 1990s.[84] In fact, after GASP pioneered the rights of citizen intervenors outside of "direct interest," the Pennsylvania air pollution law was rewritten to include specific language allowing citizen suits against pollution sources. Act 245 was passed in the 1971–1972 legislative session, and codified on the state level the specific rights for citizen intervention previously outlined by Broughton on the local level.[85]

GASP's demands for inclusion transformed the model of exclusive consensus-building in Allegheny County, something not lost on observers in the national press: "Political compromise on air pollution now seems impossible," mourned *Business Week* in 1969. Article XVII "wrests from industry the responsibility for cleaning up the pollution that remains and places that power in the hands of the commissioners. . . . This action follows five months of emotional lobbying by citizen's groups." The 1970s were going to be different, predicted these journalists: Article XVII, and Broughton's interpretation of it, had made "the appeals process not a quiet, closed-door affair, but a sort of public adversary proceeding." The *Wall Street Journal* admitted that the previous arrangement had been a sweetheart deal: "In 1960 the county and local steel companies negotiated a 'gentleman's agreement' setting a 10-year timetable for cleaning up the mills." For the 1970s, the decision allowing substantive public involvement in the local policy process meant that the days of gentleman's agreements were over.[86]

• • •

With Broughton's decision to allow more significant citizen involvement in the day-to-day process of air pollution control, the era of exclusive, elite-driven, voluntary regulation came to an end. Where previous governing bodies on the local and state levels had cultivated the appearance of public involvement without supporting substantive inclusion, new policymaking would be forced, legally and legislatively, to include representatives of the public as well as industry. The resulting negotiations, accusations, and debate were confrontational rather than cooperative; but they were also inclusive, more representative of public concerns, and possibly resulting in meaningful regulation and reform.

Legislation enabling public input did not necessitate a confrontational mode of air pollution control. Rather, confrontation stemmed from GASP's self-conscious status as a policy "outsider." Much of the rhetoric of the organization, and of the modern environmental movement, depended on a depiction of industry as exclusive and resistant to citizen involvement. Industry actually assisted GASP in this depiction by working to exclude GASP from policy deliberations. Still, if the stakeholders in air pollution control policy could manage to find areas of agreement, the new regime of pluralistic and inclusive policymaking and enforcement might provide a new opportunity for substantive improvements in Pittsburgh's air quality.

Though representatives of the public now had the right to substantive inclusion in air pollution control, it remained to be seen if citizens could create an organization to effectively exercise that right. GASP's success or failure would depend on its ability to attract, motivate, and organize citizen activists. The group's demographic makeup, and the skills and viewpoints of group members, deserve special attention.

3 "I Belong Here!"

CITIZEN ENVIRONMENTALISM IN PITTSBURGH AND THE UNITED STATES

I belong here! If you believe that, then you can walk right up in the corridor of power, on the 60th floor of the U.S. Steel building and say, look, this is my air; you don't own it because you sit in this oak-paneled office. This is my air, my daughter's air, my husband's air, my son's air, my offspring's air, and you have absolutely no right to take it as your own.

—Unnamed GASP member, from the film *I Belong Here!*

In 1975, GASP made a remarkable short film, summarizing the organization's history to that date and urging national viewers to follow its example. In the opening of the film, Pittsburgh's polluted past is quickly established with a montage of historical photographs. The pictures and music build to a crescendo when the timeline arrives at the present. The images imply activism and protest, reinforced by the rattle and crack of the marching drums on the soundtrack. The narrator, calmly defiant, declares: "In 1969 a handful of people shared a strong feeling that they had a right to clean air. These few were moved to assert that right." Immediately after the title card—"I Belong Here!" was intended as a rallying cry for any citizens who believed themselves shut out of political affairs—the establishing shot zooms in from the evening sky to a warmly lit window, bracketed by the leafy trees of Pittsburgh's relatively affluent, middle-class Squirrel Hill neighborhood. The window of GASP founder Michelle Madoff's home fills the screen, and after a cheat cut to the interior, the viewer is suddenly in a comfortable, carpeted living room filled with GASP activists. Arrayed around the fireplace and sofas sit bespectacled, bearded men and skirt-wearing young women, all engaged in earnest if not cacophonous conversation. There's a woman in a turtleneck on the couch, gesturing broadly and laughing with another woman in horned-rimmed glasses. Seated on the floor, a younger woman with long, straight hair listens and smiles. The camera pans across the graciously decorated living room, showing packed couches and overflow seating of straight-backed dining room chairs, all occupied by comfortably dressed men and women laughing and animatedly talking over one another. "Who were the people who originated GASP?" asks the narrator, and an unnamed GASP member responds: "There were housewives, union people and students; out of the original 43, probably only half of these were active. A lot of these were university people; professional people of one type or another."[1]

The film goes on to outline the group's committee structure, rhetoric, challenges, and successes, all with the intention of providing a blueprint for other activist or watchdog groups nationwide. Funded in part by an EPA grant and distributed to groups in many states, the film may have done just that. But the opening scene of *I Belong Here!* also offers clues to a number of interesting themes, most likely unintended by the filmmakers. First, where did these activists in Michelle Madoff's living room come from? GASP claimed that they "came from all walks of life" and argued that they represented all of Pittsburgh's people, but was that rhetoric or reality? How important were the "professional people" to the group? And since GASP claimed to both represent similar developments nationwide and to stimulate the formation of other activist groups, those same questions should be asked of those groups as well: Where did the new wave of local environmental activists come from? How did they organize themselves, and what arguments helped them gain the attention of the public and the authorities? Who were the citizen environmentalists?[2]

In brief, they were mixed-gender, middle-class professionals who used the language of citizenship, professional expertise, and maternalism to gain access to environmental policymaking and enforcement institutions. Their social capital afforded them entrée. As professionals and academics happy to work with existing institutions, they spent more time in committee meetings and hearings than in street protest. They were, in some ways, policy outsiders but political insiders. As white, middle-class professionals they were already familiar fixtures of civil society, even though environmental concerns were a new addition to policymaking. These environmentalists shared certain similarities with Progressive Era civic organizations, enjoying connections with existing civic associations and networks and deploying the language of expertise and maternalism. Most importantly, they were concerned with *local* environmental issues and political institutions, doing their organizing below the level of the larger, older, national groups.

This chapter examines the formation and demographics of both GASP and similar groups nationwide using internal sources from GASP, the concepts of representative membership and social capital, and organizational and institutional records for the nation. In doing so, it rescues the historical significance of local environmental organizing. In defense of the much-maligned "NIMBY" groups, interesting things happened in local and neighborhood organizing in the early 1970s.

The People and Organization of GASP

The volunteers in Madoff's living room were a part of the new, national environmental movement, although few used that label in 1969. Early observers struggled to characterize the membership of the group. The *Christian Science Monitor* found the odd conglomeration of members noteworthy, observing that GASP had enjoyed "a tremendous response from people in all walks of life—housewives, students, businessmen, professional people. And from groups-churches, colleges, unions, garden clubs." *Business Week* noted the involvement of labor but also emphasized the number of academics involved: "Conservationists and professors of industrial health, many in their 30s, provided citizen groups with technical expertise. Even the United Steelworkers, which until last year showed surprising ambivalence toward pollution control, became militant." The *Wall Street Journal* struggled to come up with a name for the phenomenon, describing it as a coalition of different groups formed to support a single interest, or "a so-called breather's lobby." But this was actually the modern environmental movement itself, uniting diverse constituencies under the aegis of environmental concern.[3]

A quick outline of GASP's first five years is in order before a more complete analysis of its internal structure, rhetoric, and status as an example of a nationwide movement. GASP was organized from the citizens who appeared to speak at the public hearings on proposed changes to air pollution control policy in the fall of 1969. Coached by the League of Women Voters, these individuals coalesced into a local citizens group complete with monthly board meetings, nonprofit tax status, a newsletter, committees, a speakers bureau, and an active and outspoken leadership whose words were featured in the local press. After organizing public testimony for the hearings of 1969, the first major battle of GASP was to gain substantive access to the workings of county air pollution control entities, especially the advisory and variance boards. From that point on, GASP leadership was visible in any discussion of local air pollution control in the public sphere; GASP speakers and educational volunteers worked to raise the visibility of air pollution issues in the city and beyond; GASP technical experts were actively involved in implementation and enforcement at Variance Board hearings; and when that body was less than fully successful, GASP lawyers were a part of the interminable legal battles that ensued.

Driving these developments was the fact that GASP was quite successful in its first five years in attracting members, keeping a core of active volunteers and leaders, and gaining attention both locally and nationally. GASP spokesmen became omnipresent in the local press. Stories about air pollution control and GASP appeared in the local press constantly throughout the early 1970s, so much so that a critical reviewer of the local PBS affiliate's new, hour-long nightly news program recommended that coverage of GASP "could be reduced to once-weekly mentions." Nationally, the EPA singled out GASP for particular mention in multiple publications; internationally, GASP was promoted to the UN as an example of successful environmental organizing and to the Soviet Union as successful democracy; Michelle Madoff was named Woman of the Year in 1971 by the local Kiwanis; a GASP representative was invited to speak to the President's Committee on Health Education in Washington, D.C., in 1972; the organization received awards from local groups including the Pennsylvania Public Health Association in 1976, the Western Pennsylvania Public Health Council in 1977, and the Allegheny County Medical Society in 1979; an editorial in the *Pittsburgh Post-Gazette* on the occasion of the group's tenth anniversary declared that it had "an unusual and lasting impact."[4] American Motors Corporation singled it out as one of four preeminent environmental organizations in the country in 1971, the same year that the Allegheny County Soil and Water Conservation District made a similar commendation. Michelle Madoff herself was named to the county advisory committee and the state Department of Environmental Protection's advisory committee. GASP itself probably summed these accolades up best; in 1975 it claimed that GASP was "one of the most successful local action groups in the country."[5]

In the film *I Belong Here!* the narrator seeks an explanation for this success, and asks, "What was a major facet in gaining the attention of the media?" An unnamed GASPer responds: "A spokesman with impact. The spokesman really raises your news potential if they are somewhat flamboyant and thrive on attention." Though this speaker might be trying to cloak the fact in hypothetical language, they were in fact talking about Michelle Madoff, who in truth was more than *somewhat* flamboyant. Any discussion of GASP must include the story of Madoff, renowned in Pittsburgh as an outspoken environmental activist in the early 1970s and as an equally outspoken member of the city council for fifteen years, beginning in 1978. Early depictions of GASP centered on Madoff and her reputation as a "practical" housewife, bringing order to city affairs. Several early articles described Madoff as a magnet for other interested persons. Her flair for the dramatic, undeniable skill at producing excellent newspaper copy, and physical attributes made her a natural for the role. "Michelle Madoff delivers her one-liners with a smile," said one journalistic portrait. "Though fluent in conversation, she is not at all formidable looking with her curly gray hair and flashing dark eyes." Journalist John Hoerr referenced the comic strip *Peanuts* to capture her defining trait: "Madoff is possessed of a furious energy," he wrote. "It swirls around her as visibly as the eddies of dust that trail Pigpen wherever he goes." GASP member Esther Kitzes called Madoff "the dynamic perpetual-motion chairman" in a *Forum* story, and Bernard Bloom remembers her as "uncontainable." In the many other newspaper stories about her activism, Madoff often portrayed her interest in air pollution in a very personal manner, related to her physical proximity to pollution, her own health issues, and her children: "Well, my husband is a cardiovascular thoracic surgeon and I'm asthmatic, and I never had a sick day in my life until I hit Pittsburgh. Many days I'd go outside to my side door . . . and you'd think you were going to choke with the taste of this heavy air."[6]

Madoff lived in Squirrel Hill, a tree-lined residential neighborhood populated by white, middle-class, and professional Pittsburghers, a heritage that both helped and hindered Madoff's efforts. As Madoff noted in an interview, "There were going to be these hearings and people could come and testify. I didn't want to go and testify and be branded as another idiot housewife—hysterical Squirrel Hill housewife in tennis shoes, as we're referred to—you know, uninformed, emotional public."[7] Describing the scene at a September 1969 seminar intended to prepare concerned citizens for the upcoming county hearings, the *Christian Science Monitor* noted how Madoff's outspoken nature helped draw like-minded people: "After the seminars adjourned, about a dozen people from the audience approached her expressing a desire to join her 'group' and help fight pollution. She did not have a 'group.' They gave her their names and addresses." At an organizational meeting of these individuals at her Squirrel Hill home, Madoff was elected president of the newly formed GASP—leading her to later joke, self-effacingly, "[m]y advice . . . is never have the first meeting in your own home."[8]

But Madoff was only one of the many strong personalities making up the diverse membership of GASP, a group of people deserving of more specific analysis. While the press described GASP's coalition as including unionists, housewives, and academics, this reflected the rhetoric of the group more than it did the actual membership. It is true that GASP occasionally enjoyed some relations with local unions, but the description of a working-class membership reflected wishful thinking rather than actual dues-paying members. Although GASP kept in contact with union locals through mass mailings, union members did not, in fact, make up a large part of the group's leadership or membership. GASP did have good relations with USWA Local 1557 president and occasional GASP board member Daniel Hannan in the early 1970s, "but that one didn't come easily," says Madoff, and relations with the steelworkers waned as economic conditions deteriorated. In reality, the other "walks of life" identified in the press coverage—"ladies' garden clubs" and conservationists and professors—were much more instrumental in GASP's creation and actions, as middle-class women, professionals, and academics formed the organizational and institutional core of the group.[9]

While the group's social status and networking shaped its strategies, the geographical placement of the group and its activities likewise determined group characteristics. Squirrel Hill and Oakland were home to research institutions, higher education, and a socially active Jewish community and upper-middle class residential homes. "Squirrel Hill was different," recalls GASP member Bernie Bloom. It "was a real community, tightest community I ever lived in. People knew each other." The majority of GASP's membership and activities were located around Squirrel Hill, in the center of Pittsburgh's many ethnically, racially, and class-segregated neighborhoods.[10] An address list from the first full board of directors meeting in February of 1970 demonstrates the concentration: thirty-three of sixty-nine attendees lived in Squirrel Hill, and fifteen more lived in the neighborhoods geographically surrounding it—Oakland, Wilkinsburg, Homewood, Shadyside, Hazelwood, and East Liberty. Put another way, 70 percent of GASP's leadership at this meeting hailed from neighborhoods representing only 18 percent of the city. GASP's first office demonstrates its connection to the middle-class neighborhoods surrounding Squirrel Hill and its location in the network of social, religious, and civic groups in Pittsburgh; the group secured offices in space provided by the Eastminster Presbyterian Church in East Liberty, north of Squirrel Hill, and stayed there for the next thirty-five years.[11]

But Squirrel Hill and Oakland were also directly downwind of the J&L Hazelwood plant on Second Avenue. In the late 1960s at Carnegie-Mellon and the University of Pittsburgh, "you couldn't think about local matters without thinking about J&L Steel—it was right over the hill," recalls Bernie Bloom. "Because, every Saturday night, it seems" the plant would do an open-air quench of slag, "generating hydrogen sulfide and the smell of rotten egg odor in the east end of Pittsburgh, so Squirrel Hill, Oakland, Point Breeze, Shadyside." As Historian Roy Lubove observed, in the 1960s "certain sections of the city, like Squirrel Hill and especially Hazelwood, still suffer at times from a thick, malodorous smog and a quick settling layer of black dust."[12] Michelle Madoff certainly experienced poor air quality in Squirrel Hill, and traced her rising environmental consciousness to the proximity of industrial and residential sites on display (Map 3.1) as her family moved into Pittsburgh in 1960:

> *We were driving into Pittsburgh; I couldn't believe the ugliness of the steel mills. Because you go down the parkway—you have to, to get to Squirrel Hill or wherever you're going. I was so depressed. I couldn't believe what I was seeing. I came from cities that had some beauty and even if there were high buildings, they were clean. You go right past the gravel pit. You know what I'm talking about: The Parkway. And the Boulevard of the Allies. Oh, was I depressed.*[13]

Madoff also emphasized Squirrel Hill's proximity to industrial sites as instrumental to her growing involvement in air pollution matters in 1969. Recalling a meeting with a young reporter from the local PBS station, Madoff describes her own medical condition as well as her home's relative proximity to the Jones & Laughlin coke works in Hazelwood:

> *[T]he reporter was John Hoerr, and John came out as a trial, and I didn't know that he had no film in the camera. And he said, "What made you*

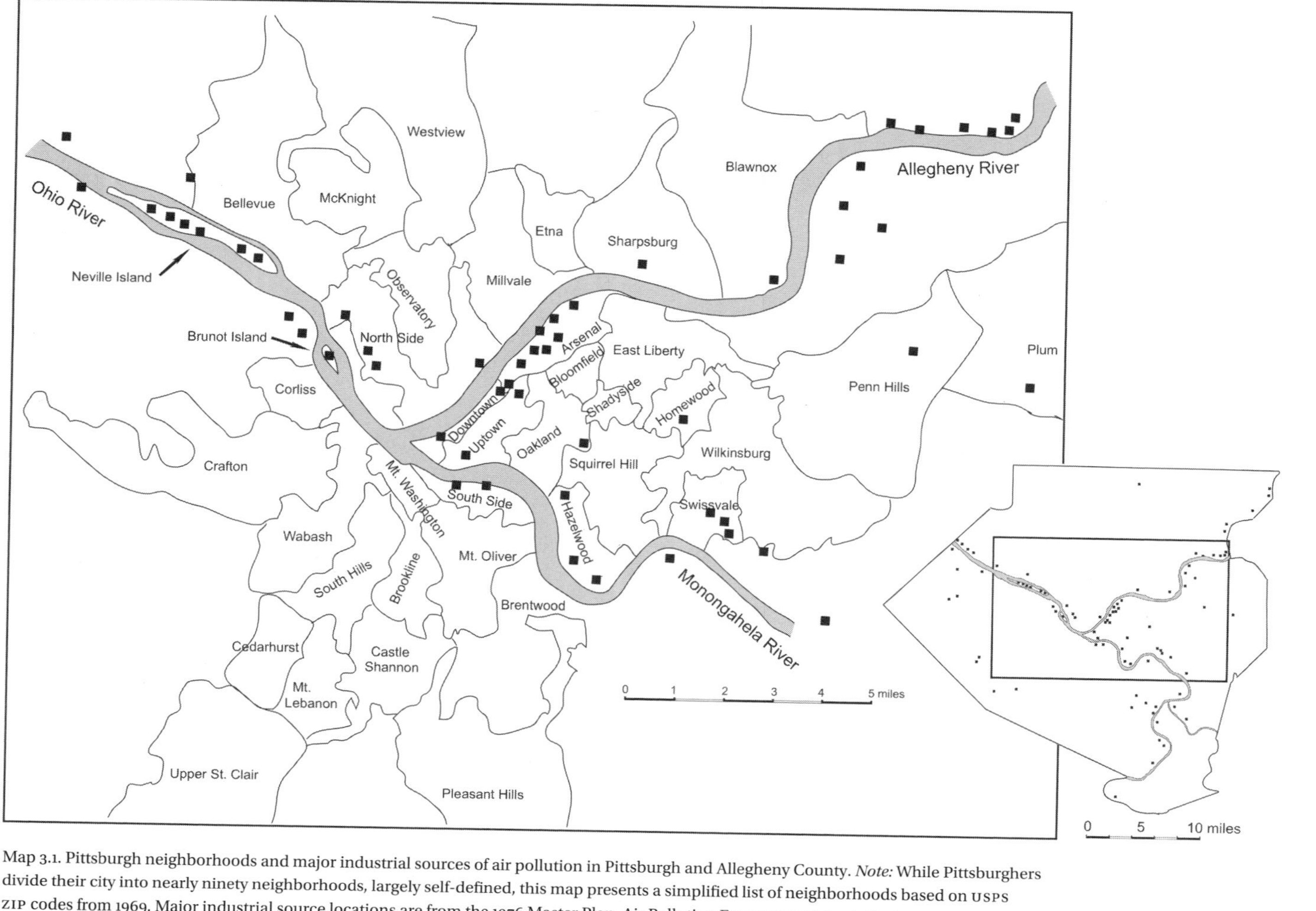

Map 3.1. Pittsburgh neighborhoods and major industrial sources of air pollution in Pittsburgh and Allegheny County. *Note:* While Pittsburghers divide their city into nearly ninety neighborhoods, largely self-defined, this map presents a simplified list of neighborhoods based on USPS ZIP codes from 1969. Major industrial source locations are from the 1976 Master Plan, Air Pollution Emergency Episode System, ACHD BAPC. Location estimated by street address.

interested in this?" and I said, "Well I have asthma and I've been in the hospital more than out, and when I look across the road, J&L is missing, because it was smoggy and you couldn't see it." He called in and said, "I think we have a story; send film."[14]

While no industrial concerns (and few if any industrial workers) were located in the neighborhood, Squirrel Hill was still relatively close to the long corridor of the Monongahela River valley, the location of some of the most heavily developed and polluted industrial areas in the United States. Squirrel Hill and surrounding neighborhoods suffered from industry's negative externalities without enjoying any direct economic benefit from industrial processes, making the neighborhood politically ripe for environmental activism. Madoff was perhaps the best example of this: her house on Mt. Royal Road was in a leafy, residential cul-de-sac that nonetheless overlooked the J&L slag heaps in Nine Mile Run.[15]

The organization's location in Squirrel Hill offered ready access to Pittsburgh's professionals and academics, but also represented a physical separation from black communities and working-class neighborhoods. This separation is evident not only in the absence of minority or working-class members but also by an absence of any perceivable attempt to make contact with these communities. Analyzing the addresses to which GASP mailed public relations material illustrates GASP's place in the network of Pittsburgh's civic society. While labor union locals, press, university, government, and ethnic groups were all well represented in this list of public relations contacts, no evidence of any contact with African American community groups exists. This omission is even more telling when compared to the inclusion of twenty-nine separate ethnic groups in the Pittsburgh area, spanning the gamut of Pittsburgh's immigrant community. The list features Eastern European immigrant groups including the Association of Lithuanian Workers, the Croatian Citizens Social Club, and the Bulgaro-Macedonian Beneficial Association. It also includes the Greek Catholic Union, the American Jewish Committee, the Irish Center of Pittsburgh, and the Polish Falcons.[16]

GASP's failure to reach out to minority groups, along with the group's successes in attracting members and networking with other institutions, can be explained with the use of the concept of *social capital,* as drawn from the work of political scientist Robert Putnam. As described in the Preface, social capital refers to an individual's ability to create and use personal and political connections within the community. GASP's male members were in position to utilize the considerable prestige they enjoyed as upper-middle-class professionals. But it was the female leaders of the organization who enabled GASP to contact what Putnam calls the "dense network of reciprocal social relations" represented by the city's many woman-centered social, civic, and reform organizations.[17] In other words, both the men and the women who joined GASP possessed a valuable resource in their access to existing, though distinct, social networks and institutions.

There was another factor behind GASP's success at finding committed, passionate volunteers: the tumultuous social activism of the 1960s. Despite the fact that Pittsburgh was not at the forefront of the social activism of the decade, there was still a long list of events in the city that politicized individuals who became part of GASP. In Oakland, there were the Black Construction Coalition protests against discriminatory union hiring at University of Pittsburgh and Carnegie-Mellon University building projects in 1969 and 1970. Downtown, the Gulf Action Project protested against Gulf Oil's corporate headquarters. There were marches in support of Cesar Chavez's grape boycott (including a march on the now-removed Manchester Bridge). Perhaps most importantly, there was an extremely active chapter of the National Organization of Women, producing multiple national presidents and a lawsuit against the *Pittsburgh Press* that went, successfully, all the way to the Supreme Court. All of this took place against a background of neighborhood organizing and community unrest. "It was very frothy, very intense period, from really 1969, 70, 71—very intense," recalls Bernie Bloom. "What I'm telling you is that there was at the same time that [GASP] was going on in Pittsburgh, there was activism happening at the very same time . . . the Gulf Action Project . . . environmental teach-in stuff in Pittsburgh . . . relationships between labor and academics and non-working class people. . . . [T]hat's the times, that's the matrix that surrounds the events."[18]

Wherever they came from and whatever their inspirations for activism, the diverse GASP members organized themselves into a group with a hierarchical structure, simple bylaws, and tax-exempt status with the Internal Revenue Service. The first bylaws of the organization, drafted and approved in the spring of 1970, named it "Group Against Smog and Pollution, Inc.," and laid out the mission: "To promote, encourage and work for the enhancement of the quality of our environment through the coordination of the creative ideas, manpower, and financial resources of conservation-minded individuals and organizations' educational, advisory and consultative efforts relative to the treatment and/or prevention of all forms of pollution on our land, air and water."[19] With that mission, GASP divided itself internally by function, as described in the film *I Belong Here!*: "We took this cross-section of volunteers and organized them into committees," says the voice of an unseen and unnamed GASPer, as the image of a manual typewriter bangs out the names—Executive, Legal, Scientific. "We used this [committee system] to come across as very technical and knowledgeable," says the volunteer. "We developed an ability to stay focused, wait out delays, and take legal action where appropriate."

The early rhetoric of GASP was moderate and accommodating, and it emphasized the reasonableness of many corporations. "We think that the air pollution control people are acting in good faith," said Madoff in the *Press*, "but we'd like to get more support behind them." GASP singled out cooperative firms for special praise, lauding several in the press for "outstanding contributions and good faith in the battle for better pollution control" and creating special certificates in their honor. Noncooperative firms got a more satirical treatment, receiving special certificates as polluters who combined to make Pittsburgh "one of the 10 dirtiest cities in the nation. . . . We salute those good neighbors who made this award possible," Michelle Madoff acerbically told the local press. But still, Madoff represented the vast majority of companies as good corporate citizens. "In pollution control, we're light years ahead of everybody else in the country. Here 95 percent of the companies that ask for variances have committed themselves to compliance and it isn't just rhetoric. Contracts have been let and equipment ordered." Even the otherwise-strident film *I Belong Here!* takes pains to avoid an all-encompassing anticorporatism. "Was industry the problem?" asks the narrator, and a GASPer responds: "Not *really* . . . ," before trailing off and placing the blame on the consumer economy more than corporations.[20]

In fact, GASP found itself in arguments with politicians nearly as often as it did with corporations. A regular opponent of GASP aims was Dr. William Hunt, one of three Allegheny County commissioners. Particularly critical of Michelle Madoff, Hunt used the rhetoric of professional expertise to counter GASP political goals: in favoring less stringent industrial regulations in the fall of 1969, Hunt told the press that he planned "a study to determine how much pollution in the Pittsburgh area comes from auto exhausts," and said that he was "planning to ask [the] County Medical Society to undertake the collection of statistics to determine if there is significant increase in illness during air pollution episodes." The journalist cut to the quick of the environmentalist argument, noting that "Dr. Hunt doubts that the effects of air pollution are as bad as proponents of strict regulations say they are." Michelle Madoff was quick to respond, countering medical expert with medical expert: "It would be a shame that a highly respected surgical colleague attempts to discredit the testimony given at the recent State hearings by local experts in respiratory disorders," she told the press. And the next day, Hunt's medical colleagues weighed in when a member of the county medical society's public health committee called Hunt's attitude "shocking" and called upon him to "reconsider his stand," going on to declare that it is the "moral obligation of the medical profession to push for adoption of new and stronger air standards." For their part, the editors of the *Press* compared Hunt's objections to "a Southern school board's replies to the 'all deliberate speed' desegregation ruling from the U.S. Supreme Court 15 years ago." By the end of the month, after it became clear that there was rising public interest in quick action on strict standards, Hunt had reversed course entirely, informing Madoff that he had been "misunderstood, and that he is strongly behind strict laws and stiff enforcement."[21]

Along with language meant to evoke a willingness to negotiate with corporations and politicians, GASP used egalitarian language appropriate to the era and supportive of women's activism. In fact, the greater the diversity of individuals involved, the more GASP was able to fulfill the ideals of its "power to the people" rhetoric. GASP publications repeatedly stressed the diversity of its membership, noting the inclusion of "blue- and white-collar workers, doctors, lawyers, businessmen, scientists, students and housewives." A GASP pamphlet put it another way: "Alone, you have just one voice. But if you gasp loud enough and with enough other people, you can do a lot. GASP represents real *people power*. We have organized into a citizen's action group to use that power." The phrase "people power" recurs in outsiders' descriptions of the group. *A Breath of Fresh Air*, the Voice of America production documenting GASP's involvement in air pollution control, describes the "people power" in Pittsburgh, noting "the underlying theme in Pittsburgh's transformation from a 'Smoky ol' Town' to a leader in the fight against air pollution is this: in a country where participatory democracy is the ideal, 'people power' has made itself felt."[22]

The phrase "people power" serves as a link from the late 1960s back to the populism of the late nineteenth century; indeed, social scientist Robert Fisher has characterized this type of organizing as a "new populist movement in the making." Fisher argues that the new populism "offered an organizing approach that toned down the late 1960s emphasis on ideology and focused on winning power and building organizations. . . . To accomplish such goals, these organizations often allied with traditional community institutions like churches, unions, and ethnic associations."[23] GASP's populist roots in traditional community organizations were enabled by the active participation of women, making them logical leaders for GASP: after all, GASP's female members provided the organizational underpinnings of the group, and the day-to-day presence in GASP offices.

Representative Membership and Social Capital

Beyond leaders and members, GASP had a third category of involvement known as representative membership, which is key to understanding the organization and its place in the air pollution control debate in Pittsburgh. From its inception, GASP followed a practice of claiming a much larger membership than it actually enjoyed. While the dues-paying membership is difficult to measure, based on budgets and receipts in the group's own records it most likely never exceeded 4,000 individuals at any point during the lifetime of the organization. Yet GASP leaders routinely claimed between 30,000 and 60,000 members in their promotional material, and declared their intention of reaching 100,000. Newspaper articles about GASP reprinted these outrageous claims without challenge. While the claim is greater than the reality by an order of magnitude, and a membership of 100,000 would have been an astounding 7 percent of the population of the entire county in 1970, it's not entirely hot air: it was based upon the inclusion in GASP's own rolls of all the members of any group that pledged support to GASP. By counting the members of the various organizations whose leaders also sat on the GASP Board of Directors, the organization could claim a much wider membership than was actually the case. GASP exploited the concept of representative membership both rhetorically, to secure a prominent place in the court of public opinion, and legally, to secure representative standing in court. This reliance on "representative membership" as a validity claim became a hallmark of the group, and is easily debunked in terms of raw numbers; but studying the groups that GASP claimed as members uncovers a very real network of social capital and gendered activism within the environmental politics of Pittsburgh.[24]

Michelle Madoff seems to have exhibited this tendency toward exaggerated membership when she attended public hearings in 1969, even before the formation of GASP. In describing her testimony at the 1969 Air Quality Act hearings, Madoff said, "I went to the hearings and testified very briefly representing the Citizens for a Cleaner Environment—a name I pulled out of the air. I said, 'Right now we have eight thousand people and you can be assured this time next year we'll have eighty thousand.' You know—I was talking into the wind." The name, the emphasis on citizenship, and the membership numbers were intended to portray a large, stable, and highly organized environmental group, though no such group ever existed. This pattern continued as GASP was coalescing: describing the scene at the September 1969 seminar intended to prepare concerned citizens for the upcoming county hearings, the *Christian Science Monitor* noted how Madoff's outspoken nature helped draw like-minded people, and how Madoff capitalized on the appeal, making contact with "a union leader with a local membership of approximately 4,000." After the subsequent organizational meeting, Madoff and GASP were quick to capitalize on that number: GASP listed its membership as being in excess of 4,000, even though only 43 had attended that first meeting. So, technically, it now claimed a membership of 4,043.[25]

The actual membership of GASP is difficult to determine, as no authoritative membership rolls are available from archival sources. But it is certainly possible to establish an upper boundary of dues-paying members by examining GASP's annual budgets and tax records. For example, in the fiscal years 1971 and 1972, GASP reported "gross dues and assessments from members and affiliates" to the IRS of $16,047.96 and $6,059.75, respectively. Assuming that all dues-paying members contributed at the lowest possible level, GASP could have had a maximum of 8,000 members in 1971 and 3,000 in 1972. With a more logical assumption of an average membership payment of $5 at the "general membership" level, GASP's reported taxable dues could only represent dues-paying membership levels of 3,206 and 1,212 for the two years.[26] These membership totals are far below published claims.

Still, GASP's claim to represent a large percentage of the population of Allegheny County was absolutely crucial to bolster its arguments for inclusion in county air pollution control processes, state hearings, and legal proceedings. While "citizen-standing" suits of the early 1970s did not necessarily depend upon a legally established minimum number of affected citizens, GASP could (and did) argue that its large representative membership qualified it, more than any other organization, to stand for the public at large. As presented in the film *I Belong Here!* GASP needed to claim wide representation in order to get the attention of political leaders: "We needed to convince the politicians that we spoke for the people," says an unnamed GASPer in the film. Madoff admits to strategic exaggeration:

> *I don't think we ever had more than 10,000 members. But we had contacts with the medical profession and scientific community and they were members. But they represented, in what they could do for us, to be a hundred people. And I remember one time that GASP was going to remove membership names from the list because they hadn't paid dues. And I . . . I said we have to have strength of numbers. You don't take any name off. And I would get to a point with people and say do you know we have 40,000 eyes looking at you. Meaning there are 20,000 people. They thought it was 40,000 people.*[27]

Creating a list of groups affiliated with GASP in its first five years cuts through the rhetoric to demonstrate the types of organizations and community networks through which GASP operated on a day-to-day basis. For example, while GASP's public relations efforts attempted to reach a large number of labor organizations (representing more than a quarter of all groups to whom GASP sent press releases), labor organizations made up only 7 percent of the total number of organizations GASP claimed to represent. GASP hoped to represent, communicate with, educate, or inform labor organizations, government, and industry. But the representative membership of the group was not overwhelmingly made up of those interests. Instead, it consisted primarily of educational, environmental, religious, and women's organizations.[28]

The list in Table 3.1 shows GASP's strong connections with a dense network of educational, religious, and environmental organizations in the Pittsburgh area. Fully a quarter of the representative membership was related in some way to educational institutions. A sampling includes the Montour High School Ecology Club (in McKees Rocks, outside of Pittsburgh), the McGibney-Lafferty PTA, the Highland School Library, the Sigma Chi Epsilon Sorority of Allegheny County Community College, and the Brentwood High School Future Nurses Club. Religious groups included entire congregations (Ames United Methodist Church, St. Mark's Episcopal Church, Bethesda Lutheran Church) as well as social or civic outreach committees from specific churches (Women's Association of Eastminster United Presbyterian Church, Women's Alliance of First Unification Church, Rodef Shalom Youth Group, Social Action Committee of Temple David). Environmental groups in the Pittsburgh area included both local chapters of national organizations (the

Table 3.1 ***Characteristics of GASP's Representative Membership Organizations***

	Number	Percentage
Groups with one characteristic		
African American	0	0.0
Civic	14	5.3
Educational	55	20.7
Environmental	32	12.0
Ethnic	0	0.0
Garden	14	5.3
Labor	72	0.6
Medical	10	3.8
Religious	40	15.0
Unclassified	17	6.4
Women	20	7.5
Youth	13	4.9
Groups with two characteristics		
Civic / women	2	0.8
Civic / youth	1	0.4
Educational / environmental	10	3.8
Educational / medical	2	0.7
Educational / religious	2	0.7
Educational / women	1	0.4
Environmental / medical	1	0.4
Environmental / religious	3	1.1
Medical / women	3	1.1
Religious / women	12	4.5
Religious / youth	4	1.5
Women / youth	3	1.1
	$N = 266$	100%

Note: Compiled from five separate lists of representative membership organizations. See note 28 to Chapter 3.

Pennsylvania Chapter of the Sierra Club) and Pittsburgh-based umbrella groups (Allegheny County Environmental Coalition).[29]

A closer look at the list of representative organizations also reveals the predominance of groups representing women, or groups that might have a special interest for women. A total of 41 of the 266 organizations specifically identified themselves as representing women, either as their sole characteristic or in combination with others, such as religion or education. Fourteen more groups identified themselves as garden clubs, quite likely predominantly female in demographic makeup, and listing female contacts or officers. Finally, a large proportion of the organizations included in the "Educational" category, itself the largest piece of the pie, were made up primarily of women, including elementary school PTAs, elementary and secondary school classes with female teachers listed as contacts, and high school science, environmental, or ecology clubs listing women as faculty advisors. From the Seeders and Weeders Garden Club, to the Plum Junior Women's Club, to the Women's Fellowship of the Birmingham United Church of Christ and the Women's Auxiliary to the Allegheny County Medical Society, GASP enjoyed clear and diverse connections to women's organizations in the public sphere.[30]

This pattern is not limited to GASP, or indeed to Pittsburgh. An analysis of the groups and organizations who attended the public hearings on local and state implementation plans for air pollution control mentioned in Chapter 1 yields similar results. In addition to the many individuals who identified themselves as "ordinary citizens" or "housewives," most speakers at SIP hearings in Pennsylvania were not linked to environmental groups but were rather associated with religious, women's, civic, or educational organizations. Mrs. Jack Matthews, chair of the Land Resources Committee of the Countryside Garden Club, attended a 1969 hearing in Harrisburg to assert that "Most of us are mothers who are asking, 'If this is what the foliage in our nature preserves, roadsides, and backyards looks like, how do our children's lungs look?' " From the Saxonburg District Women's Club to the Pittsburgh Section of the National Council of Jewish Women to the Committee on Conservation of the Fox Chapel Joint Garden Clubs, speakers at public hearings demonstrated an interlocking network of environmental, civic, and women's groups.[31]

In the words of Robert Putnam, this "dense network of reciprocal social relationships" provided the social capital that was leveraged for environmental goals, and therefore the unseen foundation of GASP's success as an organization in Pittsburgh. In particular, the horizontal connections to groups that were *not* primarily environmental in character grounded GASP in existing, and entirely noncontroversial, regional organizations. Before ruffling feathers at public meetings and county hearings, GASP laid its organizational groundwork through network-strengthening interactions with PTAs, garden clubs, and church groups. This outreach created a population of Pittsburghers interested in reading GASP coverage in the local press, likely to speak approvingly of GASP actions and concerns to neighbors and politicians, and generally supportive of environmental goals. Connections to existing groups gave GASP access to institutional resources, including volunteers for fundraising projects, donations of office space and supplies from churches and social groups, and the ability to make personal connections between institutions. Analyzing GASP's "representative membership" reveals this social capital, which might otherwise go unseen. After all, the political debates and environmental philosophy of GASP appeared on the front page of the local press, while announcements of the Seeders and Weeders Garden Club appeared on the inside pages of the "Local" section, if they appeared at all. While those activities might have been buried deep in the paper, and might individually appear inconsequential, it was GASP's ties with home-economics classes, Girl Scout troops, and church groups that enabled its visibility and vitality in Pittsburgh's civil society.[32]

Increases in Citizen Environmentalism across the Nation

Looking at the demographics of GASP's membership and its relationships with other organizations in Pittsburgh can answer some questions about the origination and identity of the newly emerging citizen environmentalists. But the argument of this work is that GASP was one instance of a broad phenomenon; that throughout the United States, a wave of citizens groups on the local level differentiated themselves from previous organizations and attempted to influence environmental policymaking. Defending this argument requires examining the many GASP-like groups that popped up across the nation. This demands a new evidentiary base, as the description of GASP activism presented so far in this chapter is based on the group's internal records and journalistic coverage. For groups across the nation, the raw numbers and names of newly formed citizens groups from directories and archival sources will have to substitute.

Most histories of the modern environmental movement in the United States begin with the observation that there was an increase in America's interest in environmental matters in the late 1960s and early 1970s. This observation is usually based on some measurement of increased interest, either through public opinion polling, journalistic coverage, political events, popularity of books on environmental concerns, the existence and increased membership of national environmental groups, or some combination of all of the above. This focus on national numbers, necessitated by the complexity of finding local groups, may have skewed perceptions. As researchers Baird Straughan and Tom Pollak put it, in emphasizing a "core of high-profile organizations, many of them national in scope," observers have created a condition where those high-profile organizations "have sometimes been conflated with the U.S. environmental movement as a whole."[33]

But while many historians have implied or inferred that there was a similar change on the local level, actual measurements of small, local environmental groups are much more rare. This is partly because such claims are contested and difficult to defend, beyond a general assertion that such groups existed and increased in number. The bare truth is that many of these small groups are and were ephemeral, coming and going in response to perceived crisis and opportunity, and rarely leaving documentary evidence of their existence. When all that was necessary for their formation was a phone number, a name, and some letterhead, these groups leave little for historians to find.

Still, there are ways to measure the existence of these groups, from increased applications for tax-exempt status, to listings in directories of charitable organizations, to correspondence with federal agencies and participation in state-level public hearings.[34] The number of small, charitable organizations with a social, civic, or environmental focus registering for nonprofit tax status with the IRS clearly increased. IRS records depicted in Chart 3.1 show that just under 2 million groups applied for exemption from federal taxes over the course of the twentieth century, with three distinct surges: during World War II, during the 1960s and 1970s, and nearer the end of the century. Adjusting Chart 3.2 to reflect increasing national population further shows the importance of the increased rate of applications in the 1960s and 1970s, with a noticeable upturn in new organizations per 100,000 population around 1970. The rate of increase in new 501(c)(3) organizations in the 1960s is greater than at any other time period after World War II. In other words, if the entire century were split into five-year increments beginning with 1901, the five-year periods beginning in 1961 and 1966 would be ranked first and second out of all such periods in the percentage each increased. The interesting thing here is the increase in organizing during the long 1960s, visible in all of the graphs in this chapter. While the impressive spike in organizing after the 1980s is what has drawn the critical attention of political scientists, there still is something happening around 1970, a spike that precedes the later, possibly inflated numbers of lobbying organizations.[35]

Although not all organizations identified the specific goals of their group in their applications for nonprofit status, Chart 3.3 shows the increase in the total number of applications per year from organizations that identified themselves as environmental or conservationist in nature. While the number of applications from environmental groups later in the century far outstripped the number applying during the decades of the 1960s and 1970s, the change in the rate of applications during the long 1960s marks a departure from previous decades. Although the envi-

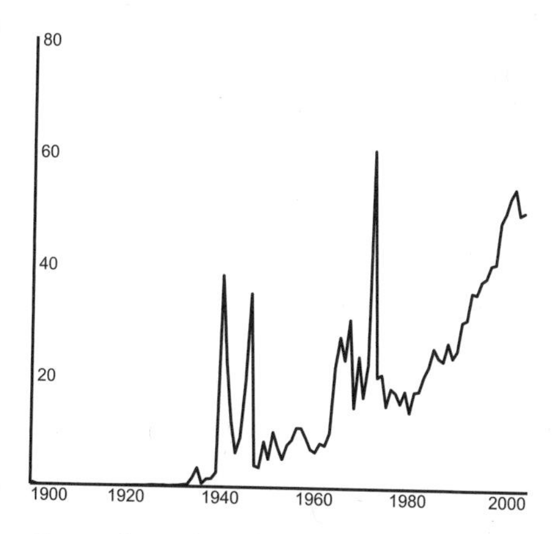

Chart 3.1. Organizations applying for tax-exempt status, in tens of thousands.

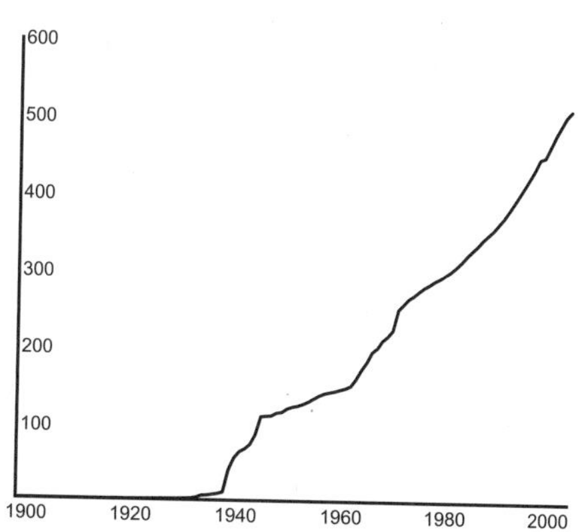

Chart 3.2. Organizations applying for tax-exempt status, per 100,000 population.

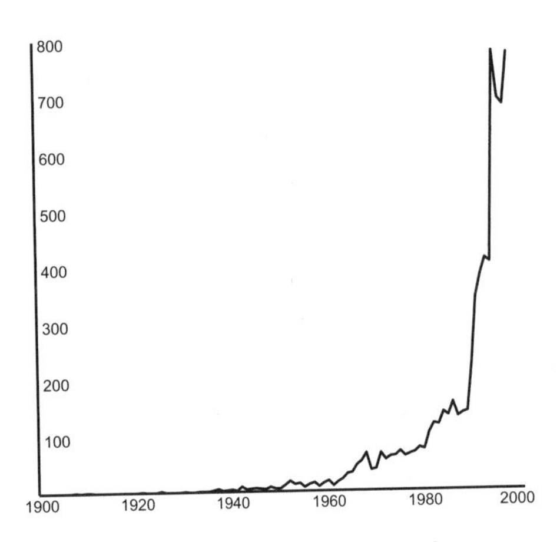

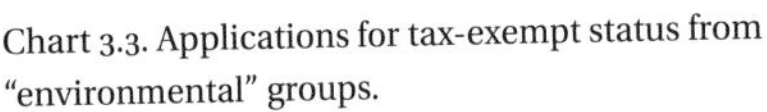
Chart 3.3. Applications for tax-exempt status from "environmental" groups.

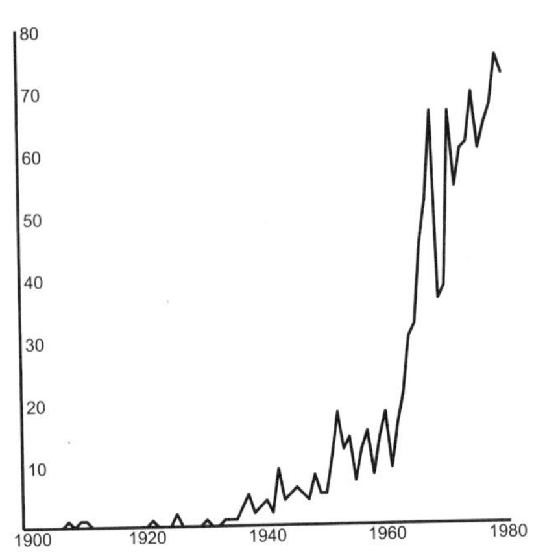

Chart 3.4. Applications for tax-exempt status from "environmental" groups, to 1980.

ronmental groups in this population are statistically few, their numbers do clearly climb throughout the decades of the 1960s and 1970s, a fact made more obvious when the phenomenal surges after 1980 are excluded, as in Chart 3.4. Before the gains later in the century, the 1960s saw a steady increase in new environmental organizations, with each year of the decade seeing more new organizations than the previous year.

Independent organizations, or groups unaffiliated with a larger national body, also were on the march. Chart 3.5 shows that while affiliated groups increased during World War II and again during the long 1960s, the number of applications from such groups fell to a murmur in the later decades of the century. The transition point is at the same time as the number of applications from independent, unaffiliated local groups rose during the late 1960s, and they have climbed ever since. This indicates that the surge in environmental activism was composed largely of a diffuse body of small, local organizations, as compared to the model of centralized national groups with local chapters made popular in the Progressive Era.[36]

Other sources support the broad conclusions drawn from the IRS data. A similar increase in nonprofit applications, and in applications from environmental groups, appears in state-level data from Pennsylvania. Applications and renewals of charitable status with the Pennsylvania Department of State Commission on Charitable Organizations skyrocketed throughout the 1960s, from a low of 2 initial applications in 1963 to a high of 111 in 1968. In fact, the 1968 total makes up nearly a third of the total applicants from the first ten years of environmental applications for charitable status. But that total of 354 applications over ten years pales in comparison to the 1,435 initial applications and renewals filed in the three-year period from 1975 to 1978.[37] This increase in charitable organizations registering for nonprofit status is corroborated by the IRS data itself, which records a jump from just over 6,000 applications from Pennsylvania groups in the 1960s to just under 10,000 applications in the 1970s.[38] Clearly, the number of state-level or smaller nonprofit environmental groups expanded, coincidental to the rise of the environmental movements, but also to the rights revolution, Bicentennial celebration, and general support for participatory democracy evident in the decade.

Analysis of the *Encyclopedia of Associations* also shows an increased number of civically engaged groups. This directory of nonprofit groups has been updated biennially since 1956, and is a common resource for sociologists and political scientists.[39] Entries in the directory are compiled from surveys filled out by the organizations themselves, or by researchers who contact the organizations, which might make the source a more accurate accounting of groups whose existence extended beyond the paperwork of a tax-exempt application. Chart 3.6 corroborates an increase in the number of all charitable groups and the number of groups coded as "civic and social." The increase of the 1960s and 1970s is still visible.

Once again, the key finding from the *Encyclopedia* is an increase in the rate of creation of new organizations. Compared to all other five-year increments, the half-decade from 1966 to 1970 saw the second largest increase in new organizations in the entire century. With more than a 41 percent increase over the number of new groups created in the previous five-year period, the late 1960s lagged behind only the five-year period immediately after World War II in the rate of increase of new charitable organizations. Put another way, the late 1960s saw the second greatest increase in the twentieth century. That finding is supported by similar analysis of the rate of increase in organizations coded as having a "Civic and Social" purpose in

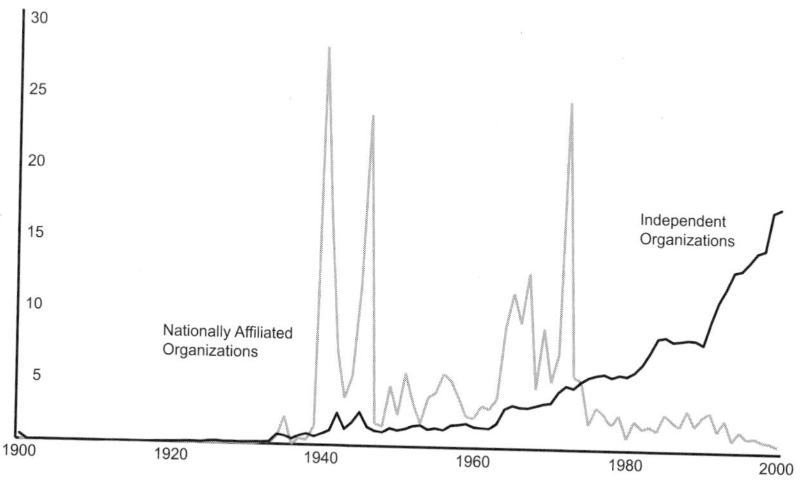

Chart 3.5. Applications for tax-exempt status from affiliated and independent organizations, per 100,000 population.

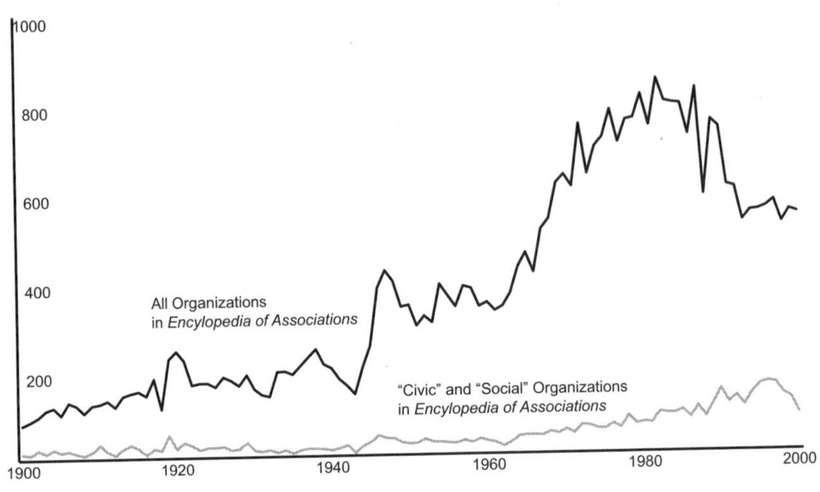

Chart 3.6. All organizations listed in *Encyclopedia of Associations*, and civic and social organizations, by founding date.

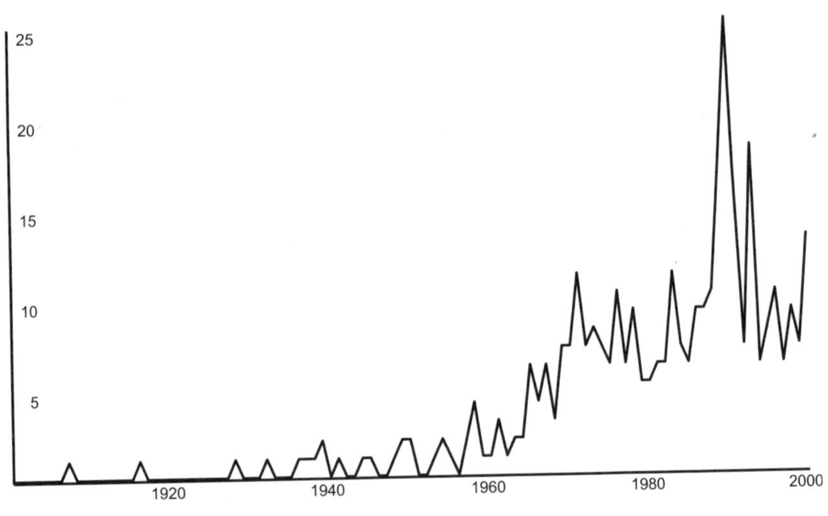

Chart 3.7. Environmental organizations in *Encyclopedia of Associations*.

the *Encyclopedia*. The late 1960s increased nearly 53 percent over the previous half-decade, and again ranked second in the century after the post–World War II period. Thus, even without correcting for the increased population later in the century, the late 1960s saw a greater increase in organizational creation than any later period. The per capita rate of new organization creation also demonstrates that the late 1960s and early 1970s saw a significant surge in organizing, again ranking second in the entire century with a 33 percent increase over the previous five-year period.

As with the IRS data above, the total population of charitable organizations here stands in for the smaller subset of environmental groups, and demonstrates an increase in organizing that most likely amplifies similar but more difficult-to-discern trends in environmental organizing. While the raw numbers of self-identified "environmental" organizations listed in the *Encyclopedia* remains too small to reliably analyze rates of change or to compare by population, Chart 3.7 still shows a considerable spike in the late 1960s. Something was clearly going on, even before the phenomenal increases at the end of the century.[40]

In Defense of NIMBY: On the Subject of Names

Occasionally, all that historians are left with is the name. Many organizations were too small or short-lived to make it into the IRS database; these "kitchen table" groups, as Baird Straughan and Tom Pollak call them, left little for historians to find. Documents from federal regulatory agencies can flesh out the statistics from the IRS, offering further records of the names, geographical locations, and actions of the new groups. Federal agencies responsible for air pollution control received thousands of letters from hundreds of state and local environmental organizations throughout the long 1960s. In some instances, these letters are the most accessible centralized record of the "kitchen table" groups. This historical record—particularly the names of the groups—encodes rich meaning.[41]

Many small groups wrote to the National Air Pollution Control Agency and its successor, the Environmental Protection Agency, in the late 1960s and early 1970s. In 1968, the NAPCA succeeded the National Center for Air Pollution Control (1966) and its predecessor, the Division of Air Pollution (1960), as the lead federal entity charged with investigating and occasionally regulating air pollution in the nation. All three of these bodies had existed under the control of the Public Health Service, itself a division of the Department of Health, Education, and Welfare.[42] The NAPCA's mission statement particularly emphasized promoting action on the state level, specifically involving the agency in advising, overseeing, and encouraging states.[43]

As a part of that mission of influencing state-level action, the NAPCA's Washington, D.C., office corresponded with countless citizens groups attempting to intervene in the creation of local or state air pollution laws. One collection of these documents received by the NAPCA and its predecessors from 1966 through 1970 shows the wide variety of these organizations. Letters from at least 174 separate citizens organizations appear in the NAPCA's archival record; many groups were frequent correspondents. Many were local or state chapters of national organizations, like the Michigan State AFL-CIO, the Metropolitan Detroit AFL-CIO, and the Air Pollution Committee of the San Diego Sierra Club. Local chapters of the League of Women Voters contacted the NAPCA from states as varied as Nevada, North Carolina, and Maine, as well as Puerto Rico. The names of other groups indicated specific health or medical interests, including the Potomac and the Hawaii Tuberculosis and Respiratory Disease Associations, or the Michigan Cancer Foundation. A great many other group names reflected their identities as regional antipollution organizations. These included the Florida Council for Clean Air, the Regional Plan Association of Southern California, the Connecticut Air Conservation Committee, and Arizonans in Defense of the Environment. Still other groups came up with inventive and original names that did not reflect their region, but whose broad geographical range nonetheless demonstrates the nationwide appeal of air pollution control. These included Citizens Against Pollution (Portsmouth, Virginia), Citizens for a Better Environment (San Antonio, Texas), the Garden Center of Greater Cleveland (Cleveland, Ohio), St. Clair County Pollution Probe (Port Huron, Michigan), SOAP (Society on Anti-Pollution, Inc., Youngstown, Ohio), Metro Clean Air Committee (Minneapolis, Minnesota), CAN (Clean Air Now, Riverside California), the Citizens Council for Clean Air (Indianapolis, Indiana), and the Better Air Coalition (Baltimore, Maryland).[44] The flood of letters from citizens groups was so overwhelming, and judged to be so central to the NAPCA's mission, that administrators eventually directed regional offices to keep a master list of citizens groups and furnish the constantly updated list to the Washington, D.C. office in order to better facilitate communication between the local and federal levels.[45]

Groups continued to correspond with the EPA's Office of Air Programs after that organization took over from the NAPCA in 1970.[46] The Office of Air Programs, or OAP, was one of many EPA offices tasked to take over the programs, personnel, standards, and practices of a wide variety of predecessor organizations.[47] More than seventy different citizens organizations from all over the nation wrote to the OAP in 1971 and 1972 alone. Their names were as varied as their locales, and included the Clean Air Committee, Life of the Land, Student Environmental Research Center, and the Northside Environmental Action Committee. Many groups emphasized their claims to citizenship with names such as Citizens for Clean Air and Water or the Citizens Environmental Council. Others indicated the theme of citizenship in explanations of their group's mission and identity. The Tri-State Air Committee noted in its letter that it was "a non-profit, citizens' organization deeply concerned about the contamination of our precious air resource," while the Eco-Coalition offered its tagline, "Citizen Action in Ecology."[48]

Some of these groups may have been inspired to shape their name, language, or methods after GASP's example. Like the NAPCA before it, the EPA used GASP as a model of this sort of effective citizen intervention, devoting three pages to discussing GASP as a successful example in a pamphlet entitled "Citizen Action Can Get Results" in August 1972. After discussing the examples of public involvement from Chicago, Pittsburgh, and San Francisco, the pamphlet's authors concluded that "citizen involvement and action has become an integral part of the movement for environmental quality in the United States."[49] GASP itself took full advantage of the opportunities to capitalize on the dialectical relationship. A letter from GASP president Pat Newman in May 1972 urged the EPA to continue programs designed to encourage citizen participation, based upon the effectiveness of previous public involvement: "I want to tell you that it is my strong feeling that the federal government realized a very strong return on its investment in the citizen participation programs," wrote Newman. "You should be proud that your agency was instrumental in providing support to an outstanding group of public citizens. . . . I have confidence that EPA will continue providing support to public citizens across the country."[50]

The names chosen by these groups reveal their members' perceptions of their identity and mission. A dominant theme in naming was the use of clever or whimsical acronyms containing environmentalist meaning, which served a number of purposes: emphasizing a focus on citizen activism, linking the groups to similarly named civil rights and neighborhood organizing organizations, and distinguishing the local activists who chose such whimsical names from more staid and traditional organizations with more institutional monikers. Previous generations of social organizations had been more likely to include the word "Foundation," "Association," or "League" in their name to indicate their affiliation with national organizations, and possibly to evoke a perception of institutional stability and respectability. But many groups in the new generation of citizen environmentalists chose to refer to themselves by short, often onomatopoetic acronyms. The acronyms spelled out names that were occasionally funny, but which often made reference to the evident goals or action-minded tone of the groups. COPE, CARP, GASP, CHOKE, CAP, and CAN were all names of small citizens groups that wrote to federal agencies in the early 1970s. The *New York Times* linked the naming trope to the new surge in environmental consciousness in February 1970, noting that grassroots support for laws was "coming from citizen groups like Chicago's SAVE (Society Against Violence to the Environment) and Pittsburgh's GASP . . . that are proliferating across the nation."[51]

Many of these acronyms began with the letter "C" for a reason: they identified themselves as "Community" or "Citizen's" groups, and encoded that status into the first word of their acronym names. From the Citizen's Clean Air Committee in Dayton, Ohio, to the Citizens Against Pollution of Portsmouth, Virginia, and Citizens Council for Clean Air of Indianapolis, Indiana, to Community Action to Reduce Pollution of Gary, Indiana, to the Citizen's Air Conservation Association and the Citizen's Environmental Task Force, both of Pittsburgh, these organizations took special pains to identify themselves in a manner that laid claim to an individual's rights in a democratic society.[52] Based upon the IRS data, environmental groups were twice as likely as all other charitable groups formed throughout the twentieth century to include the word *citizen* in their name.[53] Early press coverage of GASP itself included the word *citizen* as a prefix to the name, capitalized as if it were a silent "C" before the acronym GASP: "A proposed plan to deal with air pollution

. . . will be discussed at a general meeting of the Citizen's Group Against Smog and Pollution (GASP)," wrote one journalist in 1971.[54]

GASP was a popular acronym for many citizens air pollution control groups across the nation, though it often stood for different words, as in the Groups Against Smoke Pollution or Group Against Smoking Pollution or Group Against Smog and Pollution or evens Gals Against Smog and Pollution. People in Birmingham, Alabama, Missoula, Montana, and El Paso, Texas, founded organizations called GASP after the creation of Pittsburgh's GASP, but Washington, D.C., and Los Angeles had groups by that name before Pittsburghers independently coined the acronym for their own group. This proliferation of similarly named groups demonstrates the nationwide rise of citizens' activism channeled through self-consciously small community organizations. The mere fact that independently named GASPs were appearing across the nation indicated that the individuals who named these groups seemed to be aware of the expanding list of acronym-named groups across the nation, or at least that the acronym-naming trope had a shared meaning.[55]

Michelle Madoff argued that the acronym-naming pattern was particularly tailored to getting press coverage in Pittsburgh. After years of involvement with GASP, she left active leadership of that group to found a new organization, originally known as the "Citizens" Environmental Task Force, with the word *citizens* originally in quotes as a means of emphasis. But the name proved to be too unwieldy and obscure without a meaningful or evocative acronym. So the CETF was renamed Enviro-SOS. "What happened was that we forgot that you cannot call yourself something that does not fit into a headline of one column," she recalled. "GASP is one word. Enviro-SOS can be put in a headline. The other name didn't really work because you had to explain what it was."[56]

Pittsburgh's GASP itself spawned some of the many local acronym groups, including the Mon-Valley GASP, a spin-off organization known as GRIP or Group Recycling in Pittsburgh, and more recently, CHOC (pronounced "choke" as an adjunct to GASP) for Citizens Helping Our Community, a short-lived organization based in Pittsburgh's Hazelwood neighborhood and opposed to the revival of Jones & Laughlin Steel's previously abandoned coke works. This naming trend was obvious to observers in Pittsburgh; an editorial remarked that the name GASP originally "sounded either like a joke or a too-clever acronym." Another journalist, lamenting that there were too few progressive politicians available to run for elected office, complained that "too many natural candidates for such roles have been siphoned off into GASP, CRUNCH, WAP, FLINCH, and DUCK. Or whatever all these acronym-happy, blinker wearing groups call themselves." The joke isn't bad, and it successfully highlights the acronym-naming trend, including the tendency toward evocatively active acronyms. But the complaint also highlights a component of the definition of the citizen environmentalists: they operated outside the confines of political parties and electoral politics. Perhaps this is in fact why the acronym-naming trope existed. In order to quickly identify themselves and their mission to journalists and the public, citizen environmentalists who were not associated with long-standing political or civic institutions developed names that referenced an immediately recognizable style and contained their own markers of tone, scope, and intent.[57]

The acronym-naming trope was nationwide, and has come to be a recognizable identifier of this type and scale of political expression. No matter the city, today it is fairly likely that every daily newspaper in the United States contains a story about a local political debate that includes a comment from a group of organized citizens. It is even more likely that group has a whimsical, onomatopoetic name. That name is recognizable as a marker of small-scale, local, neighborhood activism, possibly associated with the organizational efforts of middle-class women, but not directly connected to a political party. The name is likely a clever acronym, spelling out an active word in four or five letters. If that group is commenting on land use or planning issues, the words others used to categorize or criticize the group are themselves fanciful acronyms: they are NIMBY (Not in My Backyard) groups protesting LULUs (Locally Unwanted Land Uses). Pejorative terms for this sort of activism mock the acronym pattern, which in itself denotes the recognizable importance of the acronym trope: BANANA is an acronym for Build Absolutely Nothing Anywhere Near Anything; calling a group CAVE people serves as a criticism of Citizens Against Virtually Everything.[58] NIMBY nicknames are pejorative, and "grassroots" is an unexamined positive, creating a need for a usefully descriptive replacement. If the phrase "citizen environmentalists" isn't sufficient, it might be better to call them CANCANs, for Citizen-Acronym Named Community And Neighborhood groups.

The naming pattern is so obvious that it is followed by groups formed to exploit the powerful rhetorical weapons of citizenship-based grassroots organizing. These "Astroturf" organizations were formed by corporations without real community support, and appropriated the naming trope themselves to run "false-flag" operations. As one reviewer of a book about Astroturf groups said, these groups "even resort to portraying themselves as environmentalists. One need look no further than the names (and corresponding acronyms) they choose for themselves; for example, Coalition for a Reasonable Environment (CARE) is a group whose membership includes investors, lawyers, developers, and builders." From the Citizens for a Sound Economy in 1995 (CSE) to the Committee to Preserve American Security and Sovereignty (COMPASS), to the Save Our Species Alliance (SOSA) in 2005, imitation is the sincerest sign that the acronym pattern has meaning as a claim of legitimacy, even if all claims do not contain the same amounts of truthiness.[59]

This naming pattern began with the civil rights groups of the 1950s and activism of the 1960s, and inspired the citizen environmentalists of the 1970s. Before the environmentalists appropriated it, the naming pattern included groups as diverse and contradictory as CORE (Congress of Racial Equality, the influential civil rights group founded in 1942) and ROAR (Restore Our Alienated Rights, Boston's anti-busing group of 1974). The pattern is so durable that, four decades since GASP's creation, and four thousand miles away, the implications and associations evoked by the name BARD (or Better, Accessible, Responsible Development) of England's Stratford-upon-Avon are immediately clear. Seventy years after CORE's founding, groups looking to self-identify with a certain type of locally organized citizens' activism still choose this same naming trope.[60]

Further analysis of group names indicates that while some groups were affiliated with larger, older federations and clubs, those that were not often chose to emphasize citizenship as a theme and to make themselves distinct from the federations of the nineteenth century by choosing acronym names. In this way, a new generation of local citizens groups emphasized that they had descended from the groups of the civil rights and community organizing movements—from CORE, SNCC, PUSH, MOBE, and ACORN, not the Sierra Club, the General Federation of Women's Clubs, the Izaak Walton League, the Tuberculosis League, or the American Lung Association.[61] This is particularly obvious in some examples such as St. Louis's recursively named ACTION (Action Committee to Improve Opportunities for Negroes), which was a CORE spin-off particularly focused on local matters. According to historian Clarence Lang, studying groups like ACTION explicitly leads to paying "greater attention to indigenous, unaffiliated groupings. . . . [this approach] also gives greater weight to local struggles than to the initiatives of the federal government."[62]

The acronym-named groups were and are amazingly prevalent among local and community-based environmental organizations. Groups that applied to the IRS for tax-exempt status from across the nation have chosen the pattern repeatedly since the 1970s. Some of these names are remarkably clever. There's ACORN, (A Community Organized to Restore Nature, not to be confused with the *other* ACORN), RECOVER (Regional Coalition for Ohio Valley Environmental Restoration), CURE (Citizens Urging the Rescue of the Environment), PRICE (People's Rights in a Clean Environment), STOP (Surfers Tired of Pollution), SALT (Sandhills Area Land Trust), and TRAIN (Transboundary Research and Information Program). *Clean* and variations thereupon occur numerous times, from CLEAN (Citizen's Local Environmental Action Network) to CLEAN (Citizen's League for Environmental Action Now) to the more recent CLEANUP (Citizens Learning Everything about Amoco Negligence and Underground Pollution). From ACE (Alliance for a Clean Environment) to ZOO (Zoo Outreach Organization), the naming pattern looks to be a recognizable signal of community-based environmental organizing.[63]

While this study emphasizes small, local citizen groups, it's also true that there are acronym-named groups that either rose to national prominence or were founded with the intent of garnering national attention. An evocative acronym-named group—GOO, or Get the Oil Out—was at the center of the outraged local reaction to the Santa Barbara oil spill in 1969, one of the many disasters in the decade of the 1960s that historians have fingered as seminal in creating a popular ecological consciousness. One such group became famous for generations of law students when the Supreme Court case of *United States v. Students Challenging Regulatory Agency Procedures* (or SCRAP) of 1973 set an important precedent for citizen-standing suits. In 1978, a community-level crisis sparked the creation of a citizens organization, and the Love Canal Homeowners association was born. Its most energetic leader, Lois Marie Gibbs, later went on to found the Citizens Clearinghouse for Hazardous Waste—later the Center for Health, Environment and Justice, which presented itself as a nationwide resource center to encourage grassroots environmental organizing.[64]

There is at least one more noticeable pattern. Though more difficult to quantify, the preponderance of environmental groups that received nonprofit status featured names that included their cities, regions, or areas. From the Bee County Wildlife Co-op to the Grand Mesa Citizens Alliance or the Friends Of Mirror Lake State Park, these names identified local political and civic allegiance in much the same way as words such as *citizen, community,* and *public.* They identified the organization as a component of the public sphere in that locality. Out of the more than 11,000 names of environmental groups available in the IRS data, a large share—perhaps one-third—of these groups include specific localities in their names. These include names such as the Citizens to Save Sandy Beach, Inc., the Dishman Hills Natural Area Association, and the Takshanuk Watershed Council. These place-names also indicated that these groups were small, local, bound to their immediate surroundings, and unaffiliated with some larger national organization. In other words, they were citizen environmentalists.[65]

Clearly, from the quantitative evidence, there was an increase in the numbers of such organizations. And from the qualitative evidence, these were actual groups of American citizens, not simply fund-raising lobbyists based in Washington, D.C. Those groups who wrote in to federal agencies or applied for tax-exempt status were geographically diverse, and were politically active enough to write to federal agencies. At least some of the new organizations that filed for nonprofit status or registered with the *Encyclopedia of Associations* thus truly existed, and were not the late-century "associations without members" feared by some observers. Even if later growth in organizations after 1980 was due to the creation of the false-front, lobbying, or "Astroturf" organizations described by Robert Putnam, the organizing surge of the 1960s and 1970s was genuine, reflective of many small, independent organizations created nationwide. Locally rooted social, civic, charitable, and environmental groups increased both in number and in their rate of creation in the decades of the 1960s and 1970s. GASP was not alone. The group's naming pattern, emphasis on the language of citizenship, and local focus was part of a larger movement, planting a dense network of grassroots organizations across the nation.[66]

• • •

Back in the GASP film *I Belong Here!* the final scenes leave an interesting clue to understanding the organization. The last few shots of the film show an idealized vision of a citizens group, with a symbolic leader proudly leading a ragtag group of smiling nuns, teens, children, women, and men in a triumphant procession. A marching band plays as the activists frolic on a wide expanse of green grass. This is Schenley Park, near the Squirrel Hill neighborhood, adjacent to the Carnegie-Mellon campus and looking out over the Oakland neighborhood. A woman's voice provides the explanation for the film's title: "I belong here!" she says with feeling, in a quotation that appeared as the epigraph for this chapter. "If you believe that, then you can walk right up in the corridor of power, on the 60th floor of the U.S. Steel building and say, look, this is my air; you don't own it because you sit in this oak-paneled office. This is my air, my daughter's air, my husband's air, my son's air, my offspring's air, and you have absolutely no right to take it as your own." If the language of citizenship was obviously visible in the names and rhetoric of these small, local environmental organizations, then the language of maternal care and women's activism was just as prominent. Speaking from her identity as a woman and a mother, this GASP activist encoded her right to walk into the male-dominated corridors of power within her gendered identity. This maternalist language deserves a chapter of its own.

4 Mothers of Urban Skies

ENVIRONMENTAL EDUCATION AND THE RHETORIC OF WOMEN'S ACTIVISM

After all, the chief purpose of all GASP's volunteer effort is to make our community a healthier place for our kids to live and breathe.
—GASP press release, 1975

In the early 1970s, the Group Against Smog and Pollution published this description of their mission: "GASP works within the system in a responsible manner—prodding or supporting as necessary. GASP does not ask the impossible—but does demand compliance at the earliest possible moment within the state of the art of pollution control." This is a nuanced political statement, balancing claims to mainstream legitimacy with references to scientific and technical knowledge. Listen to the careful choice of language: "working within the system" describes the members as thoughtful, well-meaning, and progressive-minded citizens in a successful democracy; they're "responsible," and do "not ask the impossible," thus forestalling criticism concerning the realities of workplace economics and the importance of industrial jobs in Pittsburgh; they are "prodding or supporting" as watchdogs that might assist the current regime rather than overthrow it, and in any case would only "prod" regulators and regulated alike, presumably in the direction in which those institutions had already chosen to proceed. They are so reasonable, flexible, and free of dogma that they can bring their support or stimulus "as necessary." Their only demand is "compliance" with existing laws, and that of course is tempered by knowledge of what is possible "within the state of the art of pollution control." This text is an outstanding example of strategy, political triangulation, and carefully chosen rhetoric—and it appeared in the opening pages of a cookbook titled *Party Cookies Only*.[1]

In fact, much of GASP's fund-raising, organizing, and educational activities took place in what might be termed women's social space, through cookbooks, garden clubs, schools, and a network of women's social and civic groups. Although women provided the organizational backbone of GASP, often served as leaders, and steered the organization into educational missions and rhetorical expressions reflective of their goals, GASP did not identify itself specifically as a gender-segregated organization. Professional men appeared in leadership roles just as often as women. As a result, historians and political scientists have written about GASP only in terms of its apparently gender-neutral environmental politics, without prying beneath the surface and observing the explicitly maternalist rhetoric and female organizational base of the organization. Noted political scientist Charles O. Jones wrote an entire book on the politics of air pollution control in Pittsburgh during his time there in the 1970s, and his contemporaneous account describes GASP as a civic organization with scientific expertise without mentioning GASP's gendered rhetoric or educational mission at all.[2]

Many historians of the modern environmental movement—that is, the groundswell of environmental concern and activism marked by the first Earth Day in 1970 but with philosophical and organizational roots throughout the postwar era—have overlooked the importance of the gendered rhetoric of that movement. This oversight exists despite the extensive histories of civic and urban reformers of previous time periods, in particular the Progressive Era. These studies have demonstrated the importance of women's groups and activism. Maureen Flanagan, Sueellen Hoy, and Andrea Kornbluh have all explored what Kornbluh calls the "unique relationship between women and the city" evident in Progressive Era women's civic organizations.[3] Many environmental historians have examined the gendered logic underlying American history, from Virginia Scharff's edited volume *Seeing Nature through Gender* to Carolyn Merchant's groundbreaking work in *The Death of Nature* and elsewhere. And historian Adam Rome has specifically urged additional exploration of the role of women's activism within his appeal for history that places environmentalism within the context of the 1960s.[4]

But most historians of the modern environmental *movement* have not spent as much time identifying the gendered roots of activism in the 1960s and 1970s, preferring to link the rise of environmentalism with increasing concerns over population pressure, scientific and technological change, and evolving aesthetics of the leisure class. This is a mistake; and it's a mistake in interpretation that logically follows from a national focus. If one focuses on the national development of the environmental movement in postwar America, then the narrative of changing federal law, successions in environmental leadership, and nationwide economic developments explains the rise in widespread awareness of environmental issues. This narrative highlights congressional hearings, headline-grabbing politicians such as Senator Edmund Muskie and President Richard Nixon, and the connections between the first Earth Day in 1970 and the political leaders and New Left students who helped to organize it.

Urban environmental historians can offer a slightly different explanation, one that pushes back the chronology of environmentalism, one that takes into account preceding movements, and one that pays more attention to who, exactly, constitutes the environmental movement. Focusing on the city as a unit of analysis rather than the nation reveals continuity with Progressive Era civic reformers rather than the rise of New Left activism. Older, more explicitly anthropocentric public health debates take center stage instead of new environmental philosophy concerned with population bombs and ecological breaking points. The protagonists become middle-class female civic activists, who used the gendered rhetoric of maternalism and an astonishing social network of women's organizations to support environmental activism. On the level of the city, and with the tools of the social historian, the environmental movement is not a sudden transformation in need of an explanation; rather, it represents increased attention to long-standing debates over urban space and power. In cities the environmental movement was an opportunity for middle-class women to repeat their Progressive Era arguments concerning maternalist care for urban space—but at least by the 1960s, they had more access to the public sphere.

This is not to say that the modern environmental movement was a movement solely of women. But even excluding ecofeminist organizing, many mixed-gender groups still featured prominent female leaders, enjoyed implicit connections with women's social networks, were based on the organizational skills of women, and used rhetorical allusions to maternal care for the natural world. Historians recounting the stories of Rachel Carson and Lois Gibbs have employed gender as a category of analysis for the modern movement; such observations should be turned to environmental organizing where women's activism is less immediately obvious, but rather requires some investigation to demonstrate. Elizabeth Blum has already taken a step in this direction, examining the meaning of gendered activism at Love Canal underneath the compelling but distracting story of Lois Gibbs. This chapter follows a similar path, considering the language, politics, and organizing of women in both the leadership and rank and file of environmental activism in Pittsburgh.[5]

Pittsburgh and the Women of GASP

If the new generation of citizen environmentalists emphasized female care for the environment as a rhetorical tool, they did so in a city dominated by a masculine self-image. When Pittsburgh's civic activists, local leaders, and industrial representatives voluntarily came together after World War II they made some progress on long-delayed plans for smoke control. The area had long been associated with the male world of industrial work. It was "The Smoky City," the location of a sprawling complex of coke ovens, iron and steel works, and factories. The football team was named the "Steelers," and their helmets sported the corporate logo of the largest steelmaking company in the world; the local beer was named "Iron City," and the city itself went by nicknames ranging from "Steel City" to "Smoky City" to "the forge of democracy." Pittsburghers considered their civic identity to be tied up with smoke as a symbol of industrial success and continued employment. This version of civic identity continues into the twenty-first century. The recent announcement of the new Steelers' mascot, burdened with the ridiculously overcompensating name of Steely McBeam, offered a burly, masculine, and ethnic steelworker with a permanent five-o'clock shadow as a personification of a city in which there are no longer any active steel mills. Partially because of this self-identification as a productive steel town, there was little likelihood of success in controlling smoke in Pittsburgh before the end of World War II, resulting in decades of poor air quality.

All attempts at smoke or pollution control before the 1960s were negotiated through voluntary agreements of male city leaders, elites, and captains of industry. The "Pittsburgh Renaissance" was a creation of closed-door agreements between Mayor David L. Lawrence, an urban machine boss in all but name, and the Allegheny Conference on Community Development. The ACCD was exclusively male, a businessman's organization with city booster goals, boasting the membership of almost all regional business interests and power brokers. The group's most famous member was industry titan Richard King Mellon, inheritor of the Mellon banking fortune and ardent civic booster. The ACCD could offer an "advance bipartisan consensus" among business leaders in support of specific policy reforms.[6] Thus, those involved in policymaking were exclusively male, white, and generally members of the city's business elites. They often invoked a vision of a productive industrial city in discussing their civic goals, which actually included some form of pollution control. But the benefits of smoke or pollution control were related to business concerns: increased desirability of the city as a location for industrial and corporate concerns, enhanced performance of healthy workers, and improved competitive status of Pittsburgh nationally. As one member of the ACCD noted in a public speech,

> *Smoke must go. Our region, freed of smoke, slums and decay, can in our own day use the magnificent mountain air, the panoramas of our valleys, the inspiring might of our mills to make a world where the enormous energies of industry are balanced by a way of life that will hold and draw the best of labor, the best of technical skills; that will give them a place to raise their families in health and in the air around a proud and growing community.*

The reasoning behind pollution control, for the ACCD, was civic boosterism and successful industrial production. Even the raising of families was in the service of "drawing the best of labor" to the region, as a means of increasing production. The air could be cleaner, in this vision of the city, in order to better serve "the enormous energies of industry."[7] The influence of the ACCD, however, was waning. By the late 1960s, a newly pluralistic and confrontational process in many American cities replaced the previously exclusive model for policymaking. In Pittsburgh, GASP was at the forefront of this development.

The conditions surrounding the creation of GASP were arranged almost exclusively by women. After attending training sessions conducted by the League of Women Voters to encourage public participation in upcoming hearings in 1969, Michelle Madoff and two other women circulated a letter calling for a meeting at Madoff's home. Those early organizers wrote in search of a wide membership, as we've seen, addressing their letter to all "Fellow Breathers." This letter described a diverse membership, including citizens and experts alongside businessmen and "housewives." The resulting first board meeting of the organization, in February 1970, drew sixty-four attendees. Of these, twenty-seven were women who were to become permanent members and active leaders of the group, while only six male attendees went on to take leadership roles.[8]

There's some confusion over the demographics of GASP that can only be addressed through a social history approach to archival documents. National and local press coverage of GASP's efforts emphasized the group's rhetoric of diversity while highlighting the technical expertise and academic credentials of professional men. A casual reader might think that the organization was a cross-section of American race and class. In reality, the majority of GASP's activities depended upon the organizational energy, skills, and social capital of women, most often homemakers married to academic or professional men. The group's daily activities depended upon the volunteer labor of middle-class women. Four of the first six presidents of GASP were female: Michelle Madoff, Pat Newman, Ann Cardinal, and Joan Hays. The organization's sole paid employee was a female secretary. Fragmentary lists from GASP letterhead indicate that women filled an average of one-third of all leadership positions during the first five years of the group's existence. GASP's women were particularly active in membership, community-building, and educational roles, shaping the rhetoric and goals of the organization in ways that reflected a gender-based understanding both of environmental ideals and political realities.[9]

Simply being listed on group letterhead does not demonstrate active involvement, as many veterans of community organizing well know. Women's relative importance might actually be diminished when compared in this manner, since the GASP letterhead habitually listed a large number of male professionals who were almost never present at organizational meetings. For example, while the combined letterheads list 243 total names with 77 of those belonging to female members, several letterheads also list (somewhat self-importantly) the entire roster of GASP's legal or medical committees. These committees were exclusively male and were rarely in evidence at GASP meetings, and their inclusion on the letterheads skews the gender ratio toward the male. In fact, GASP women were more likely to regularly attend leadership meetings than their male counterparts. While women made up approximately a third of the leadership during the first five years, they made up two-thirds of those who attended board meetings. Put simply, women made up the majority of attendees at leadership meetings and were more than twice as likely to show up at these decision-making sessions as men who were in similar leadership positions.[10]

Many locally prominent men were listed as a part of GASP leadership but were rarely involved in the group's activities or meetings. For example, Dr. Cyril Wecht (who gained national prominence as Allegheny County's celebrated mystery-solving coroner in the 1980s and 1990s and as a target of unsuccessful and controversial federal prosecution in the 2000s) was listed as a member of the GASP Board of Directors for three consecutive years—but never attended board meetings during that period, at least according to GASP records.

Those women who did attend board meetings were mostly young, often college-educated, occasionally newcomers to Pittsburgh, and in some cases married to locally prominent men. This list begins with Michelle Madoff, but doesn't end there. Madoff was an exceptional case; and it would be incorrect to tell her story but exclude the many other activist women in GASP. In fact, there was a long list of women organizers, many more than can be included here. But even this summary demonstrates reservoirs of social capital through specific ties to Pittsburgh women's and civic organizations.

Michelle Madoff became the public face of GASP and environmental concerns in the early 1970s. Her self-identification as a housewife and mother helped to define GASP's character. Born in Toronto, Canada, as Pauline Radzinski, Madoff grew up in a foster home and went by "Millie" before legally changing her name to "Micki Rodin" after her last name proved too much for her instructors, who failed to call her name at a school assembly rather than attempt to pronounce Radzinski. She lived in Boston and New York City and worked as a dance instructor and advertising copywriter before marrying Dr. Henry R. Madoff, a heart surgeon, in 1958. After moving to Pittsburgh in 1961, Madoff worked to keep house and raise their daughter, Karen, born in 1960. She was not politically active before the fall of 1969 but became highly involved both in GASP and in the wider world of Pittsburgh politics throughout the 1970s. The local press called her, among other things, a "housewife and irate asthmatic," but Madoff identified herself in early GASP documents as "Mrs. Henry R. Madoff." Madoff's outgoing and brash personality made her a reliable source of outrageous quotes for the local press, and journalistic coverage of her thoughts made her a figurehead for the local environmental movement, to the point where she gained national attention. For example, a visiting editor for the Rochester (New York) *Democrat Chronicle*, searching for a personality symbolizing Pittsburgh in the 1970s, declared that "just maybe it's the crusading, pin-sharp wife of a surgeon, Mrs. Michelle Madoff, founder and first president and still an indefatigable spokesman for an environmental group known, to some of the nation as well as to Pittsburgh, as GASP." After leading GASP for several years, she left active leadership of the organization to create her own group, the "Citizens" Environmental Task Force, later Enviro-SOS. While still involved in local organizing and GASP even after resigning as president in 1972, she held a number of appointed posts, from a prominent position on the county's Air Pollution Control Advisory Committee beginning in 1970 to the state's Citizen Advisory Council of the Department of Environmental Resources in 1971. After years of politicking, including quixotic independent campaigns for the powerful Pittsburgh City Council in 1973 and the county council in 1975, she eventually made it to the city council in a special election in 1978, and then proceeded to serve—colorfully—for the next fifteen years.[11]

Marilyn F. Janocko, a longtime member of GASP who repeatedly served as the board secretary, was born in 1941 in Pittsburgh. Educated at the Carnegie Institute of Technology, she graduated in 1963 with a degree in art education. She was politically active in the League of Women Voters both before and during her long association with GASP, joining the League in 1969 and becoming the environmental chairman of the League's Allegheny County Council.[12]

Pat Newman was the recording secretary for GASP's board meetings before becoming president of the organization in 1971, immediately following Michelle Madoff's tenure. Often signing her name as "Mrs. John B. Newman" in the first several years of her work with GASP, Newman was a 1963 graduate of Stanford University with a B.A. in philosophy. After moving to Pittsburgh from the Pacific Northwest around 1965, Newman became quite active in a number of civic organizations, including some centered around her home in Pleasant Hills, a neighborhood in the far south of greater Pittsburgh. A 1971 profile in a local newspaper noted her affiliation with a vast list of other civic organizations, including "the LaLeche League, South Hills Association for Racial [Equality], YWCA, League of Women Voters of Mt. Lebanon, Wilderness Society, Planned Parenthood, Environmental Action, Audubon Society, Western Pennsylvania Conservancy, Pittsburgh Symphony Society, Pittsburgh Zoo Society and Trout Unlimited." Following her time in Pittsburgh, Newman returned to the Northwest, attended law school, and began practicing law.[13]

Ann Cardinal, GASP's third president, did not have Newman's civic pedigree. After moving to Pittsburgh from St. Louis, she began working with GASP in January 1970 only as an occasional break from child care: "She switched baby-sitting with a friend to spend one day a week in the GASP office to help out," noted a newspaper story. Cardinal's increased political activism and involvement seemed to surprise even her: "If someone had told me three years ago I'd be doing this," she told the press after she was named president of the group in 1973, "I wouldn't have believed it. It's basically not me." As with other GASP volunteers, her activism was located within her role as a mother. Cardinal explained her conversion to GASP with the story of a fishing trip in St. Louis. "My 5-year-old daughter, who was 3 at the time, laid back on the grass and said, 'Gee, Mom, I didn't know the sky was this color blue.'" Ann Cardinal moved from local environmental activism to working as a spokesperson for the EPA Region III in the 1980s, then into consulting as a community relations expert in the 1990s.[14]

Pat Pelkofer was one of GASP's founders, and was active in the League of Women Voters and president of the Pittsburgh Hearing Association when that group spearheaded an anti-noise pollution campaign in the early 1970s. Born in 1927 in Pittsburgh, she was an Oakland resident and a graduate of the University of Pittsburgh. She ran many of the day-to-day affairs of GASP, including editing the monthly newsletter and serving as vice president. "For quite a while in GASP she was the unpaid everything," one GASP member observed. She served as a GASP board member for an incredible twenty-nine years, and during the 1970s and 1980s was appointed to multiple state-level advisory boards.[15]

Jeannette Widom followed a slightly different path to, and within, GASP—as a native Pittsburgher (and sometimes Squirrel Hill resident) whose personal interests lay in the direction of the baking contest of the Allegheny County Fair (she took top honors for six years). But she also donated her baking and writing skills for GASP fundraisers. As special projects chairman for GASP, she wrote, promoted, and sold thousands of copies of her three cookbooks. At around three dollars each, the cookbooks raised tens of thousands of dollars for GASP projects and reached a broad, diverse audience of homemakers, women's groups, and corporations who might not otherwise have had any contact with the environmental organization. And yet her interest was not entirely baking related; her activism in antismoke protests dated back to 1946 and the voluntary reforms that the United Smoke Council urged on industry at that time. Twenty-five years later, she urged citizens to attend yet another public hearing: "I think we've waited long enough. . . . The air is no better than it was 25 years ago. One of the scientists in GASP says it's worse."[16]

Contemporaneous observers noted the presence of these motivated and effective women in the air pollution debate. *Pittsburgh Press* conservation editor Fred Jones wrote a valedictory address for Pittsburgh's women pollution fighters in 1973, arguing that local environmental organizations were to be congratulated on their professional demeanor and preparation. "Their spokesmen are intelligent, highly articulate and remain unflustered under questioning," he writes, observing that "the great majority of these individuals are women, most of them young, on the sunny side of 40. Most of them, but not all, are university graduates. A surprisingly large percentage of them are married to professional men—doctors, lawyers, research scientists, college professors and corporate officials." A 1972 story in the alternative newsweekly the *Forum* highlighted women's environmental leadership. Though overstating GASP as being made up of "9/10th female members," the article goes on to argue that "a cross section of GASP members [shows] that female environmentalists are all middle class and above average educationally." Author Josephine Schmidt observed that "women have continued to carry the ball on ecology. . . .

[W]e do the everyday day-in and day-out work of grassroots environment[al] activity in our own area as well as the rest of the country."[17]

Flour Power and GASP's Educational Mission

The impact of women's concentration on the "everyday day-in and day-out work" can be found in the educational wing of GASP's work. While the most visible component of GASP's environmental activism took place in the courts and the Variance Board hearing rooms, a great deal of time and effort on the part of the group's volunteers went into educational activities. Through a variety of outreach programs, speakers bureaus, national seminars, and documentary films, GASP attempted to teach Pittsburghers, and Americans in general, two subjects: air pollution and participatory democracy. This educational mission was carried out largely by women, and GASP's educational rhetoric was shaped *by* its female volunteers and *for* the group's largely female social networks. GASP's language generally followed the developing themes of the postwar environmental movement and included references to ecology, public activism, and population concerns. But statements from the educational components of the group also displayed a uniquely gendered view of the city, its government, and women's place in relation to both.

GASP's educational mission was funded through the production, promotion, and sale of cookbooks, a project overseen exclusively by women. The cookbook projects and baking fund-raisers were central to GASP's identity. They linked the new organization to a network of preexisting women's groups, offered positive publicity for the organization, and furnished a high profile for environmental rhetoric in Pittsburgh commercial space. They even managed to make a significant profit every year.

GASP first turned to baking as a fund-raising project in the spring of 1970. According to a GASP publication, an unnamed GASP member brought cookies to an "informal gathering" of the group. The cookies were in the shape of GASP's recently created mascot, "Dirtie Gertie, the Poor Polluted Birdie." According to a GASP mailing titled "She's Some Cookie!" the "finished birdie was two-toned, with outstretched wings sprinkled with pollution (chocolate jimmy), crossed eyes—from breathing all that bad air, and a cashew-nut beak." Under the direction of Jeannette Widom, GASP members quickly decided to sell Dirtie Gertie cookies as a fund-raiser at public information booths. The cookies were baked by volunteer labor drawn from a network of women's groups, as a GASP publication noted: "Now everybody knows that a lot of baking goes on at churches. And there are teen-age cooking clubs. The YWCA. And the Girl Scouts. Ecology clubs. Home Ec. Classes in the schools. . . . In church kitchens, at YWCA clubrooms, in a cafeteria at a girls' academy." GASP tapped into all of these preexisting women's organizations for baking duty to support its goals, claiming that groups "from junior high teens to senior citizens" contributed to bake cookies "in a true expression of community." In a March 1971 story, a small local paper illustrated the network of female religious groups that had been enlisted in support of GASP: captions on three different pictures identify women from the Emory United Methodist Church Women's Society of Christian Service, the Emory United local activities committee, Eastminster United Presbyterian Church, and Point Breeze Presbyterian Church. "Busy women . . . participated in the GASP 'Dirtie Gertie' bake off recently," says the caption, naming the volunteers, including the minister's wives from both Emory United Methodist and Eastminster Presbyterian. At least forty-two separate women's organizations appear as volunteer bakers in GASP's records. With the help of this network of religious, civic, educational, and women's groups, GASP had emulated Tom Sawyer: volunteers were lining up to complete the group's work.[18]

For the first promotional sale, twelve volunteer groups each made one hundred cookies, with donated dough from a local bakery and toppings from a supermarket. Jeannette Widom's husband used tin snips to create cookie cutters, and along with recipes and instructions, the combined ingredients and tools were distributed to volunteers in a GASP-supplied kit. The completed cookies were sold for a charitable donation of twenty-five cents each, certainly a high price for 1970. GASP's "flour power"—their phrase—raised three hundred dollars for that first round, and countless thousands thereafter, but also raised public awareness of the organization. GASP members regularly sold the cookies in the Jenkins Arcade, setting up an "Air Pollution Information Station" there. GASP continued "Fighting Pollution with a Rolling Pin," eventually claiming that the "Pittsburgh public now regards [the annual cookie sale] as a tradition."[19]

The regularly staffed Jenkins Arcade information table was an important representation of GASP in the city's public space. GASP's appearances before the Variance Board and its omnipresence in the local press were points of contact with the intangible thing known as civil society. But GASP's offices in the Eastminster Presbyterian Church and the Jenkins Arcade location were physical linkages with both civil society and with women's space. Before the arrival of suburban malls in the region, the Jenkins Arcade indoor shopping area was a centerpiece of the downtown neighborhood, a destination shopping attraction. As the Jenkins Arcade kiosk was organized by the primarily female educational wing of the group, and since it was set up during the workday, the location was staffed exclusively by GASP women and catered almost exclusively to women shoppers. These women sold gag cans of "fresh air," bumper stickers, and Dirtie Gertie cookies, at the same time handing out literature and discussing air pollution matters among the women shoppers. Many stories in the local papers summarized the outreach: "Mrs. Paul Sikar of 753 Montclair St., Greenfield, 33-year-old mother of four, said the public response to the campaign was 'amazing.' . . . 'People kept coming up and complimenting us on the work we were doing and expressing the desire to help,'" reported the paper.

Both Sikar and the journalist continued to emphasize the GASP volunteer's status as a mother caring for her family's health:

> *Mrs. Sikar said her husband, a Brizilian* [sic] *physicist now taking post-doctoral work at the University of Pittsburgh, wants to move the family to Maine to get away from polluted air. "I remember when we came to Pittsburgh from Brazil," she said. "He told us to take a good look at the sky because it would be the last time we'd be seeing a clean one for a while. He was right.*

The Jenkins Arcade table was a part of the *I Belong Here!* film as well, with an unnamed female GASPer—possibly Ann Cardinal, this time—donning the Dirtie Gertie costume and dancing in front of the table to attract the attentions of passersby. A picture of yet another female GASP volunteer, Mrs. Irving Nadelhaft, appeared in an above-the-fold, front-page picture in the *Press*, selling the cans jokingly labeled "Clean Air" at the same time that Pittsburgh was suffering from a very real air pollution emergency. A year later, the Jenkins Arcade station was still going strong, as the paper reported that "two little boys dropped a stink bomb . . . at an information stand sponsored by . . . GASP," but that "in a few minutes, under the onslaught of seven pairs of feminine feet, the sulphur dioxide fumes were dissipated," and the volunteers could get back to selling Dirtie Gertie cookies, handing out anti-DDT literature, and discussing the relationship between household purchasing and pollution.[20]

GASP members continued to sell Dirtie Gertie cookies whenever possible, but the economic and organizational importance of the cookie sales was soon eclipsed by the GASP cookbook projects. The three cookbooks written by Jeannette Widom were *Party Cookies Only* (produced in 1972 and sold at retail for $2.50), *Just Coffeecake* (1974, $3.00), and *Fun Buns for Kids* (1975, $3.00). The cookbooks expanded GASP's organizational network to encompass housewives and cooks who would not otherwise have supported or known about the environmental organization, while allowing GASP to develop corporate sponsors. *Party Cookies Only* was underwritten with a grant from longtime Pittsburgh packaged foods producer H.J. Heinz Company, and was a "year round best seller providing a steady income" to GASP, according to its author.[21] The second book was sponsored by a quartet of downtown department stores: Gimbels, Sears Roebuck, Kaufmann's, and Horne's, each of which offered $800 toward printing costs and then agreed to sell the books in their stores.[22] The three books together sold an estimated twenty thousand copies over five years, with profit from sales from 1972 to 1975 coming in at around $30,000—not bad for an organization with a total income in 1972 of only $39,000.[23]

But sales figures do not fully illustrate the impact GASP's cookbook projects had in connecting the organization to the Pittsburgh community. The cookbooks offered groups and corporations an opportunity to work with a local environmental organization in a venue that seemed noncontroversial and moderate. Opportunities to sponsor or support the cookbook projects drew corporate supporters that had no previous or subsequent involvement with GASP. For example, a party promoting the release of the third book was sponsored by a number of local corporations: Allegheny Air Lines (later US Air), Sears Roebuck, three natural gas companies (Columbia, Equitable, and People's), and the local Coca-Cola bottler. Anonymously, the Calgon Corporation, A&P food stores, and the United Steelworkers of America also contributed to the party, which featured a guest list of representatives from thirty-five local newspapers, magazines, and radio and television stations. The press party gave the natural gas companies a venue to quietly make their case for their own claims to provide a clean-burning fuel, while other corporations most likely enjoyed the association with a noncontroversial component of a local civic organization.[24]

GASP also sold quantities of the cookbooks to other volunteer and civic organizations at a reduced cost for fund-raising projects, allowing those groups to resell the books themselves and to keep the profits. The many resellers of the cookbooks are not cross-listed as GASP "representative members" but include many of the same types of organizations: religious women's groups such as the Squirrel Hill–based Beth Israel Sisterhood or female-oriented social groups like the Blackridge Garden Club. While these groups got the opportunity to raise money for themselves, GASP built social capital and spread its message throughout the ramifying networks of women's social and civic activism. Even when GASP placed its cookbooks on display for sale in the many different Pittsburgh-area businesses, it was extending its visibility into new territory: From Marshall Fields and Gimbels to Waldenbooks and Kieffer's Card Shoppe to Chez Gourmet, GASP's cookbook was placed before the Pittsburgh public.

This is especially noteworthy because while GASP cookbooks raised funds and created networking opportunities, they also disseminated an environmental message to a broad general audience. Dirtie Gertie cookies were a clever antipollution argument in themselves, coming from an organization that otherwise did not spend much time on issues of wildlife. The cookbooks likewise contained information about GASP and its air pollution control goals, mentioning the group's diverse membership and educational efforts, while noting that "[r]epresentatives of GASP, along with attorneys and scientists[,] appear on behalf of the public at all Variance Board hearings." The *Fun Buns* cookbook, which offered recipes for children, not only included a recipe for making Dirtie Gertie cookies but also a children's story about Dirtie Gertie, paired with a reproduction of the Allegheny County Air Pollution Index: "Every day [Dirtie Gertie] reads the Pollution Index in the paper or on TV," noted the accompanying text. "The Pollution Index lets everybody know whether the air is 'satisfactory' or 'unsatisfactory.' Clean or Dirty. But most times it's the same sad, sad news: 'Unsatisfactory Air.' This makes Gertie *very* unhappy."[25]

In all instances, the GASP cookbooks blurred the line between environmental advocacy and domestic improvement. Promotional material for one book was headlined: "GASP Cookbooks: Sweeten your baking—and the air too!" Another promotional pamphlet for *Party Cookies Only* defines GASP as "a nonprofit citizen's organization whose goal is to *clean up the environment* in Allegheny County." A third press release punned on the domestic and environmental goals of the cookbooks: "All proceeds from the sale of 'Fun Buns' for kids will 'raise dough' to keep GASP projects for a cleaner environment in Allegheny County alive."[26] The conflation of housework and civic reform in the language of GASP publications was, most likely, entirely intentional. GASP's blurring of domestic and municipal goals echoes the rhetoric of Progressive Era municipal housekeeping, which argued for the logical extension of women's domain from the home to the city itself. But it also reflected the experiences and lives of the women of GASP, who were likewise blurring the lines of domestic and municipal responsibility.

GASP's cookbooks, while educational in their own right, provided funding for a wide variety of more overtly educational projects, including two widely circulated documentary films, a variety of publicity-seeking stunts, a busy speakers bureau, leadership conferences, and promotion of a much-debated air pollution index for Allegheny County. In educating the public about environmental concerns as well as the power of participatory democracy, GASP demonstrated a commitment to an educational mission that, until now, has escaped the notice of observers. GASP's educational mission did not receive the high level of public attention that greeted its fight for inclusion on the Variance Board or its various legal battles with large industry in the Pittsburgh area. But the group's educational mission reached a large audience with a message that emphasized children's health, care for future generations and the citizen's right to participate in regulatory decision making.

Perhaps the most visible of all of GASP's educational projects were the organization's two widely distributed films: *Don't Hold Your Breath (Fight for It!)* from 1972, and *I Belong Here!* from 1975. Both films place equal emphasis on the themes of air pollution control and citizen involvement in the regulatory process. The first of the two films was written, filmed, directed, and edited by GASP member Esther G. Kitzes. The director of student publications at the University of Pittsburgh, Kitzes held an M.A. in educational communication, and had produced a number of instructional films before beginning her work with GASP.[27] *Don't Hold Your Breath* was originally conceived in a GASP proposal to the Office of Environmental Education of the Office of Education of the United States Department of Health, Education and Welfare (HEW) in 1971. The $72,200 grant proposal was intended to fund a series of environmental seminars and a film tentatively titled *Yes, You* Can *Fight City Hall*. The proposal argued that the film "will highlight the problems and solutions found within Allegheny County in particular to make the county's own citizens more knowledgeable; in addition, it will demonstrate how a citizens' group can play an influential role in reducing pollution in any county."[28]

GASP received only $10,000 from HEW in 1971, but continued with a scaled-back version of the project funded by cookbook and cookie sales, complete with an eighteen-minute, 16 mm documentary film.[29] *Don't Hold Your Breath* opens with a scene of elementary-school age children on a playground. Slowly, some of the children gather around a piece of chalk graffiti: "Pittsburgh stinks," one of them reads, and they agree: "Yeah, Pittsburgh *does* stink." The following scenes depict young Pittsburghers, mothers, and GASP academics offering explanations for the poor atmospheric conditions. GASP member and Duquesne University biologist Emmanual "Manny" Sillman stands on a barren hillside overlooking the United States Steel Corporation's Clairton Coke Works. "All that is left is this desolation," he says, describing the absence of any plant life. "The killer is sulfur dioxide." Subsequent scenes show Linda Compano, a registered nurse who was not a GASP member, sending her children out to the school bus wearing white surgical masks over their faces. Finally, Carnegie-Mellon economist and GASP board member Lester Lave comments on the balance between environment and employment: "As an economist, I say, do not be fooled by statements that it would cost too much, or that jobs would be lost."[30]

The film was distributed and reprinted for a number of outlets nationwide, and GASP rented it out to organizations throughout Pennsylvania. The first group to see it was at a "preview showing . . . presented to presidents and conservation chairmen of major women's organizations—civic, religious and cultural—at the [local PBS affiliate] WQED studios."[31] Perhaps the most attentive audience was in the state capital, Harrisburg, as a local newspaper noted that "members of the State House of Representatives sat through a 'dirty movie' and then heard a call for stiffer fines for polluters. The movie, shown yesterday by Mrs. Pat [Newman], president of the Pittsburgh-based Group Against Smog and Pollution (GASP), 'starred' the belching black smoke from the quenching operations at the Clairton Coke Works."[32]

The next film was *I Belong Here!* and actually resembled the film proposed for the 1971 grant application as *Yes, You* Can *Fight City Hall*. Under a new grant from the Office of Public Affairs of the Environmental Protection Agency, GASP produced a film that provided a blueprint of successful organizing and participatory democracy, while briefly mentioning environmental problems. As a GASP press release put it, "Basically, 'I Belong Here' is the GASP story—presenting in a refreshingly entertaining way—the principles, methods, and strategies which have made GASP one of the most successful local action groups in the country."[33] The film itself consisted of various scenes of GASP volunteers at work: stuffing envelopes, meeting in a private home, and testifying at a Variance Board hearing in the Gold Room of the Allegheny County Courthouse. In an artistic choice that probably made sense in 1975, scenes of actual environmental organizing in the film are prefaced by whimsical set pieces by a mime troupe, whose members attempt, through silent action, to demonstrate the intricacies of community organizing: a mime-citizen discovers himself waist deep in pollution, and wonders where it comes from. A mime-activist then unsuccessfully attempts to convert passersby with environmental slogans. Finally, a single mime-speaker sways a doubtful audience, then leads the triumphal march through Schenley Park.

But what is perhaps the most compelling image from the film is not staged by the mime troupe. Toward the end of the film, as a publicity stunt for GASP, a gaggle of five-year-old children equipped with antipollution placards floods the wood-paneled chambers of the Pittsburgh City Council. Shepherded by their mothers, the children held signs that read "I deserve to breathe clean air," "Your present is my future," and "I belong here!"[34] In fact, "I belong here!" is the rallying cry for the film, a basic statement that GASP—its majority-female membership in particular, and the American public in general—possessed an irrevocable right to an active role in regulatory decision making. The point of *I Belong Here!* is not to emphasize environmental philosophy; it is to empower citizens in order that they might act on environmental philosophy. The film spends more time talking about the committee structure of the group than it does describing Pittsburgh's pollution.

With *I Belong Here!* GASP sought to offer itself as a model for citizens groups across the nation. The film offered both an organizational blueprint and a plan for successful intervention in government affairs. As a part of the general approval of the goals of participatory democracy and citizen environmentalism, this activity was supported by federal agencies. As noted, HEW financed the creation and distribution of *Don't Hold Your Breath,* and *I Belong Here!* was funded by the EPA. The message got out; copies of the films were purchased or rented from GASP by a variety of organizations throughout the early 1970s. Groups in nineteen states rented prints of the film in 1972 alone.[35]

In addition to the films, GASP made innovative and varied efforts to draw the public's attention to both environmental ills and participatory democracy. These GASP publicity stunts included selling the previously mentioned cans of "Clean Air" (presumably shipped from outside of the region) as a fund raiser. A Massachusetts newspaper reported that the cans were priced at $1.25 each, and noted that "the directions say to merely flip off the lid and breathe deeply. Those who would rather save the contents for a choky day would be advised to keep the lid on."[36]

As a part of GASP's public outreach, the group maintained an active speakers bureau: GASP experts who volunteered to speak to any organization that invited them. The speakers bureau reached out to the community, building social capital by promoting the academic credentials and professional expertise of the GASP leadership to offer compelling speaking engagements for civic organizations and educational venues. But perhaps the most audacious of GASP's educational projects was the Leadership Seminar for Citizen Action—an EPA-funded seminar to encourage other citizens groups to emulate GASP's success. According to a GASP newsletter, "30 citizen leaders from 15 states" convened in Pittsburgh for the event in October of 1972.[37] As in *I Belong Here!* participants were trained to replicate GASP's organizational model and citizen-watchdog role in their home states.[38]

GASP mascot Dirtie Gertie featured prominently in publicity efforts, depicted on bumper stickers, shown in animated public service announcements, and even appearing in person. "Dirtie Gertie" awards were made yearly to the dirtiest polluters, handed out by a costumed, anonymous female GASP member. Wearing red long johns, flippers, and a papier-mâché head, an unidentified GASP member costumed as Dirtie Gertie was actually ejected from the Pittsburgh Hilton on February 14, 1972, after attempting to present Edwin Gott, chairman of the board of U.S. Steel, with that year's "Dirtie Gertie" award for worst polluter. It's not entirely clear who was in the costume, though Madoff is a likely culprit. At least in this case, the figure of Dirtie Gertie allowed the women of GASP to publicly repudiate the exclusively male leaders of industry. Or at least they could have, had hotel security not intervened.[39]

GASP's mascot flew haltingly through the polluted skies of Pittsburgh in three animated television public service announcements, eventually crashing to the ground because of the ill effects of the air pollution. The PSAs proclaimed that "Clean Air is for the birds, and People too!" In subsequent PSAs Dirtie Gertie gave birth to an equally dirty baby birdie: "Dirty Dick, the poor polluted chick," and Gertie was often depicted in a nest, caring for her child, in various posters and stickers.

In another wide-reaching attempt to publicize pollution, GASP organized "Pollution Land tours" of industrial sites throughout the county. GASP members organized their own tours, but also offered a self-guiding version. The brochure for the self-guided tour, subtitled "The Pity of Pittsburgh: Pollution Land," urged visitors to

> *SEE! The U.S. Steel Edgar Thompson Works! (UGH)*
> *MARVEL! At the J&L Works! (Phew)*
> *GASP! At the Mighty Clairton Works! (Retch)*
> *GAG! As Birds Die and Old People Cough!*[40]

A number of additional programs offered educational opportunities for young and old alike. Michelle Madoff used an Earth Day story in 1971 to proclaim that "Protecting the environment should not be limited to one week a year," and announce that GASP was "making environmental teaching skits available to students, teachers and ecology groups to provide information on pollution problems." There were more formal educational projects, too. "Project Environmental Rebirth," funded by the same HEW grant that supported the production of the GASP film *Don't Hold Your Breath*, brought twenty Allegheny County high school students together for a weeklong seminar on the University of Pittsburgh campus. Originally envisioned by Esther Kitzes but overseen by Mrs. William B. (Mollie) Herron, the seminar included environmental lectures, "rap sessions," a trip to a Pirates baseball game, and a tour of the Duquesne Light Company's Shippingsport Power Station.[41] In targeting youth as well as in choosing the term "rebirth" as a part of the title, GASP's women indicated their understanding of environmental education as a reproductive process.

The Language of Maternalism

GASP's educational materials took great pains to highlight the group's responsibility to the health and well-being of children. A draft form of one GASP brochure asked the reader: "What will you tell your children when they ask why they can't catch their breath while playing in the yard? . . . GASP needs your help today in order to insure your child's right to breathe tomorrow." One press release described the primacy of this goal: "After all," noted the release, "the chief purpose of all GASP's volunteer effort is to make our community a healthier place for our kids to live and breathe." In films and interviews, group members reemphasized the health of children: "Mrs. [Pat] Newman brought up that recurring image of Clairton school children wearing surgical masks to avoid inhaling particulate matter into their lungs," one interviewer noted in 1971, referring both to a scene filmed for *Don't Hold Your Breath* and an image repeatedly mentioned in discussion of pollution matters in Pittsburgh. Madoff promoted the voices of school-age children, with an entire 1970 *Press* story devoted to children's letters to GASP: "I am 10 and, if I get married when I grow up, I won't want my children to live or die in polluted air," wrote one young girl precociously using maternalist language. A twelve-year-old boy used GASP's other favorite rhetoric: "I would like to do my part as a young citizen." The paper proceeds to include Madoff's own children in the story, closing with the point that "her daughter, Karen . . . literally papers their home with 'Snoopy for Clean Air' posters." One GASP member combined her daily walk with two potent symbols: "When Mrs. [Johanna] Hicken takes her 2-year old son, Robert, for a walk in a stroller, she first pins a sign on her back reading simply 'Stop Pollution.'" Hicken appears with stroller and sign in the accompanying photo, under the odd headline "Car-Free Wife Battling Pollution in Murrysville."[42]

GASP's promotional brochures highlighted children's health, combining a conciliatory, rational claim to responsible political action with a maternal emphasis on the topic of children's health. In one section of a four-page brochure titled "Responsible Action is the Key," the author wrote, "[w]e of GASP are not out to soothe our collective conscience with grandstand acts. Our approach is informed, responsible action . . . and perseverance. We are fighting for you and your children for the right to breathe clean air." Another section is headed "OUR POOR CHILDREN," and reads:

> *We do not yet know all the long-term effects of air pollution, but we do know that we're using our children as guinea pigs. And already documented effects*

prove we are leaving them a deadly legacy. . . . On many days in Pittsburgh mothers cannot let their children outdoors to run and play. In Clairton residents have reported that their children often walk to school with noses and mouths covered with handkerchiefs to try to keep some of the bad air out of their lungs.[43]

GASP volunteers highlighted the importance of maternal responsibility for the health of children by acting out maternalist themes through the group's mascot. Soon after the cartoon "Dirtie Gertie" was created, GASP members decided that their female mascot should have a child. GASP held a contest for school-age children to name the young polluted birdie and eventually sent some of the contest entries that came from the Crippled Children's Home to the local press. A letter from a GASP volunteer to the *Press* noted:

I was touched by Luellena Davis' entry . . . who suggests the name "'Little Wheezer' because my little brother has asthma and when the air is polluted he wheezes." . . . So read the entries—you will probably laugh—and cry—as we do—cry because of what we are doing to our kids when we live in a world that pumps polluted air into their lungs.[44]

The winner of the contest was an eight-year-old boy from Mt. Lebanon, who named the baby "Dirty Dick, the Poor Polluted Chick," beating out "Polluted Peep" and "Clean Jene."[45]

Several extended versions of Dirtie Gertie's story portrayed children and their health as central to GASP's mission. One story for children describes how Dirtie Gertie flies "past the steel mills and the factories," looking for clean air:

Sometimes she takes her baby, Dirty Dick, the poor polluted chick, with her for a breath of fresh (?) air. He looks just like his mommy. And he coughs just like she does too. This worries Gertie because she knows that dirty air is not for the birds. . . . And especially not for children![46]

Another version of the Dertie Gertie story has Gertie looking for clean air in Allegheny County. This script calls for a female member of GASP to dress up in the Dirtie Gertie costume and care for a child dressed up as Dirty Dick, accompanied by an audience of school-age children providing sound effects: "Up and down the rivers she flies—past J&L in Hazelwood (cough cough). Past U.S. Steel works in Clairton (choke wheeze). Past Braddock. (Gasp.) Homestead (cough)." Without finding any fresh air to breathe, Dirtie Gertie (G.) must still care for her child, Dirty Dick (D.D.), in the polluted atmosphere. The stage directions continue:

G. disappears & immediately returns with D.D. in stroller. G. goes through business of fussing over D.D.—feeling his brow, shaking her head sadly, taking his "pollution index" and then giving him a big dose of "pollution solution" off the big spoon . . .

[Narrator:] "Poor Dick," she says — "You always have such a stuffed up head. And a runny nose. And you sneeze so much. Surely there must be clean air to breath somewhere in Allegheny county!" (G. helps Dick blow nose noisily in huge handkerchief with much tender fussing . . .)[47]

It should not be surprising that GASP's mascot was a mother. After all, the volunteers who worked in the parts of the organization using Dirtie Gertie were almost all women. Educational, fund-raising, and membership projects were conceived and carried out by women, and Dirtie Gertie symbolized their status as mothers caring for future generations in the polluted environs of the Smoky City. Dirtie Gertie was not just an animated bird but symbolically a woman, rhetorically a mother, and literally an honorary Girl Scout.[48]

Dirtie Gertie was an imagined mother, a symbolic woman whose home was the wide, polluted sky above Pittsburgh. As Vera Norwood observes, "Historically, one of American women's primary metaphors for describing plant and animal life has been as a home."[49] In describing the urban environment as a home through the eyes of an animal mother who cared for the sick and the young, GASP members tapped into a long history of successful gendered rhetoric concerning the domestic sphere, nature, and morality. When other female activists had used these rhetorical constructions, they succeeded in mobilizing existing political rationales for new battles, as Norwood notes: "When, in *Silent Spring*, Rachel Carson located the threats of chemical pollution within domestic spaces, she knew she could count on the support and activism of a generation of American women who had inherited a century-long history of female activism aimed at protecting not only human homes but the plants and animals of our larger home — the earth."[50]

Similarly, the women of GASP provided a gendered, contrarian view of the city. They described Pittsburgh itself as one large home, and evaluated the city based not upon economic success or entrepreneurial spirit, but as a healthy and safe nursery for the rearing of children. When Dirtie Gertie appeared in animated PSAs, her nest was shown perching precariously over an industrial landscape of factories and smokestacks. This was a graphic presentation of the proximity of homes and home life to the productive center of Pittsburgh.

Journalistic coverage of GASP, whether intentionally or not, always emphasized the gender of the environmental activists as much as their environmental message. Early coverage of the group invariably specified GASP activists' status as women: "Mrs. Henry R. Madoff, housewife, mother and tireless antiques collector, is also an 'irate asthmatic' and a crusader for cleaner air," said a biographical piece in the *Christian Science Monitor*. "Wife Wars against Pollution" proclaimed the headline of another story chronicling the work of Joan Hays, GASP president and leader of a coalition of environmental groups. Strangely enough, though the headline identifies Hays only as a *wife*, the text of the story never mentions her *husband*, even while quoting her extensively on air pollution matters.[51]

A local paper's biographical story on GASP president Pat Newman begins with a jarring non sequitur: "Did you know that living in a polluted atmosphere such as ours reduces the wear of nylon stockings? This actually has nothing to do with the decision of Mrs. John B. Newman of Pleasant Hills to become active in GASP . . . but it still remains a fact." Even though fashion seemed to have "nothing to do" with Newman's motivation, the topic was still deemed important enough to appear as the lead of the story. The author quickly identified Newman's actual, non–stocking related motivation, returning again to the theme of maternal care for children's health: "What really was the deciding factor in her interest in environmental conditions was the concern about the high level of DDT showing up in mother's milk. At this point, the young mother of four was truly disturbed and decided to take action." Later in the story, the journalist returns to the topic of Newman's children, especially her newborn infant: "And considering the health of her children, ages 6, 5, 2, and none months, there is a lot at stake." Toward the conclusion of the story, after listing a bewildering array of civic organizations to which Newman belonged, the story again emphasized her involvement in feminine activities: "She laughs when asked what she does with the rest of her time, 'I don't do the things most women do such as go to coffees, play bridge or get my hair done but I do find time to work in my organic garden, play the piano, sew, knit and cook.'" This journalistic coverage of an environmental activist never leaves the theme of Newman's gender; the characterization of her activism begins and ends with femininity.[52]

National press coverage likewise drew special attention to the feminine component of GASP, focusing either on the outspoken or "peppery" Michelle Madoff, or describing how "corporations . . . once scoffed at 'emotional,' uninformed citizen-environmentalists" but that GASP's competent and well-prepared female activists had battled that dismissive attitude. The characterization of environmental activists as "emotional" appears repeatedly in national coverage of Pittsburgh's air pollution debate, almost certainly as a thinly disguised dismissal of activist women. A *Business Week* piece from 1969 noted that the creation of the Variance Board "followed five months of emotional lobbying by citizens groups and conservationists." In late 1971, state Chamber of Commerce spokesman William Tipton "condemned the 'emotionalism' he said citizens groups have aroused" at public hearings, while simultaneously making the overwrought claim that proposed legislation would put a quarter of all Pennsylvania companies out of business and that industry was being "crucified" by the public. Madoff described herself as an "irate asthmatic" almost as often as others dismissed her as a "paranoid asthmatic"; either way, the phrase implied that she was unbalanced or at least motivated by emotion or mental illness.[53]

GASP members took great pains to disarm this type of attack. In response to the Chamber of Commerce spokesman who felt crucified, Pat Newman, then president of GASP, attempted to reverse Tipton's accusations, arguing "that while environmentalists are not particularly hysterical about the Implementation Plan hearing, Mr. Tipton's remarks certainly do strike us as hysterical and quite extreme." As previously mentioned, Madoff herself said that she made up a name for a fictitious environmental group before she attended her first public hearing in 1969 to avoid being labeled "another idiot housewife—hysterical Squirrel Hill housewife in tennis shoes, as we're referred to." Madoff specifically countered what she described as the image of an "uninformed, emotional public," and her public statements often included spirited defenses against characterizations of environmentalists as "emotional, uninformed or silly people."[54] GASP member Pauline Nixon made similar disclaimers in public, even after she had presented a remarkably concise and well-supported statement to a state Environmental Quality Board hearing: "I apologize if I come off as an 'upity [*sic*] female'. I am simply reflecting the feelings of the many environmentally concerned organizations who I have talked with in the past weeks preparing for today." This desire to avoid the label of hysteria is also evident when the GASP newsletter drew special attention to a "surprising profile of American environmentalists." In the words of a GASP writer, the Opinion Research Corporation reported in 1973, "by sex, environmentalists are evenly divided—50% men and 50% women. This apparently contradicts the somewhat intuitive assumption that the environmental movement in the U.S. is principally made up of 'a bunch of housewives with mimeograph machines.'"[55] GASP publications describing the group as mainstream and nonconfrontational appeared just as often as descriptions of the group's feminine nature, possibly as an attempt to defuse objections to women's involvement in the political process. The conciliatory language that begins this chapter, of "working within the system," "prodding" or "supporting" agencies and industries working for change, and claims that the group "does not ask the impossible," was surely an attempt to reassure supporters of industry that the group was rational, not emotional.[56] Madoff herself juxtaposed the desirability of a cooperative organization with an undesirable feminine "emotionalism": "We are not militant housewives, we work through the system," she told an interviewer for the Voice of America radio documentary: "I think it's more with the calm planning, with good help from the experts who are concerned" that success was possible.[57]

Later in her political career Madoff contended with local politicians, always male, who characterized her behavior in a similar manner. Madoff's adversarial relationship with Dr. William Hunt, one of the three elected county commissioners during GASP's early years, was emblematic. In 1972, Hunt called Madoff "a paranoid asthmatic" whose "emotions exceed her common sense." To be fair, Madoff had just called Hunt "a disgrace to his profession" after he criticized the Allegheny County Health Department as "alarmist." Three years later, however, when Madoff ran for Hunt's county commission seat as an independent, the war of words was back on: "She's a paranoid public nuisance," Hunt declared. "She's always coming up with some half-truths which she palms off as a great secret. It's always crap." After he lost the election, partly due to Madoff's splitting of independent voters in her failed candidacy, he was even more blunt: "there should be laws against paranoid people like Michelle running for public office." Madoff recounted these attacks frequently, and seemed to revel in them: "They call me the Housewife with a hatchet, the Voice of Venom, a gadfly, a witch with a capital B," she told a group of student journalists. While it is defensible to describe Madoff's relationship with the press as intentionally sensational and tending toward the conspiratorial, labeling Madoff as "paranoid" and "the gossip candidate" was undoubtedly an attack with overtones of gender bias. Madoff was aware of this tendency, but used it to her advantage: "You can say I'm crazy or call me neurotic or tell them I'm obnoxious . . . just so you do the story," she said in a newspaper story headlined "Candidate Likes to Make Waves: Call Madoff Names—She 'Loves It.'"[58]

In the national press, which generally spoke of Pittsburgh activism in approving terms, Madoff was almost always described in terms of traditional feminine roles: wife, mother, or homemaker. A 1970 *Christian Science Monitor* article on the group began by noting that "[t]he hands that rock the cradle are rocking industry here." It further described Michelle Madoff's children, featuring a sketch by her ten-year-old daughter showing a cartoon poodle saying, "I hate *Poluted* [*sic*] *Air*." The article emphasized Madoff's relationship with her children and husband even more than her activism for clean air, and even seemed to credit GASP's success to its female leadership: "Everyone knows that if you want something accomplished, you turn it over to a busy woman."[59]

Occasionally, the perceived need to identify GASP activists as women resulted in the use of words rarely heard in the English language, as when state EQB member John M. Elliott found the need to introduce Michelle Madoff as both "a quasi legend in Western Pennsylvania . . . and foundress of GASP." There was a degree of euphemism employed by many observers; GASP's Madoff was described alternatively as "colorful," "peppery," and "outspoken"; this was some sort of polite code for either Madoff's admittedly colorful, peppery, and outspoken personality or for the phenomenon of a politically active woman. An editorial from an out-of-state newspaper called her a "Joan of Arc type," which implied either that she was inspiring, inspired, or insane, most likely some combination of the three. At other times, no euphemism was necessary; simply identifying the activists by gender was enough to indicate that these women were outsiders in the policy process, as when one GASP publication recounted that Madoff was known simply "as 'That Woman' to some industrialists."[60]

GASP members explicitly identified themselves as mothers and homemakers in public statements, often using the identification in the same manner that male members of the group used to identify technical, business, or academic credentials. "As a representative of a majority of women known as homemakers, I feel it is my duty to speak briefly of my awareness and concern with the problem of air pollution," said Caroline Sadler to the assembled members of the state EQB. "The dynamic activism of GASP . . . roused me from apathy, educated me, and moved me to help rally the 'people power' of informed citizens," she continued.[61] After identifying herself as a "housewife" to a different public hearing, Mrs. Duvall A. Jones used a household allegory to counter arguments against pollution control:

> *Suppose I decided one day that, inflation being what it is, I had to cut costs to maintain our household budget. . . . As an economy move, I stop washing the dishes. . . . The dishes become greasier and smellier and our meals are first unpleasant, later poisonous. Could I justify all this by saying, 'It would cost more to wash the dishes, we can't afford it?' Of course not . . . the job of feeding a family includes cleaning up after each meal and maintaining an unpolluted set of dishes, pots and pans.*

This allegory recasts the responsibility of corporate polluters; if motherhood required the responsibility of both feeding and cleaning, then why didn't producers also have a responsibility for washing up after themselves?[62]

The use of motherhood to support antipollution arguments was not limited to GASP members. Under the title "Concerned Mother," a *Press* story highlighted maternal presence at the hearings out of which GASP coalesced in 1969. Under a photograph showing a young mother with a bundled infant on her lap, the caption reads: "Mrs. Bradley Segar of Shadyside showed up at today's hearings before the state air pollution commission with her 4-month old daughter, Myra, in tow. Unable to engage a babysitter, Mrs. Segar attended the hearings anyhow 'as a citizen who is concerned about the air we breathe.'" Two years later, a front page *Post-Gazette* account of another public hearing highlighted maternal identity. Beneath the headline "Mother Dislikes 'Sulfur Pot': 50 Air Strong Views at Pollution Hearing" appears a photograph of Dorlaine Ziegler holding her small child, its head on her shoulder as she speaks. The article noted that Ziegler was "one of a number of mothers who appeared cuddling babies in their arms." Both Ziegler and the headline writer were positioning motherhood as a component of antipollution protest.[63]

Local and national coverage of GASP's female leadership continuously emphasized feminine and maternal characteristics of the organization. Widely accepted images of maternal care, negative stereotypes of "emotional" testimony, and counterclaims of rational, cooperative action all reflect an understanding of the implications of prevailing gender roles in political activism. Michelle Madoff, at least, specifically understood that she was playing the role of maternal authority figure in public debates:

> *I thought that my role in being taken seriously was that I should look like a matron. I should look very serious: no high heels, no short skirts. So I went out and bought tweed suits. You got this? Tweed suits, tailored suits, and I would wear lace-up shoes. This is true! . . . then one day I was late for a session and I was in a stretch jumpsuit, that was made to look like denim but it wasn't, and it stretched and it clung and it looked great. And after that meeting [Duquesne Light] objected to my presence.*[64]

The women of GASP were clearly redefining their status as political activists. But the most obvious marker of this transformation lies in the way that GASP's female leaders identified themselves. Over the first five years of the organization's history, almost all of the prominent female leaders of the organization went from identifying themselves by their husband's name in official correspondence and newspaper coverage (Mrs. Henry R. Madoff, Mrs. Allan L. Widom, Mrs. John B. Newman) to using their given names in similar instances, often dropping the title "Mrs." entirely. This may have represented the women's growing political independence, or a cultural shift in the acceptance of female political activism, or the influence of the women's liberation movement and its consciousness of gendered language. In the case of Madoff, however, her decision to stop using her husband's name was only one step in her increasingly independent and outspoken public persona. By the time of her candidacies for local political office in the mid-1970s, she was referred to as "Ms. Madoff" in newspaper stories. She ran as an "independent Democrat" in the overwhelmingly Democratic city, seemed to delight in stirring up political trouble on the city council, and was divorced from her husband by 1981.[65]

While this analysis of GASP publications, journalistic coverage, educational efforts, and cookbook fund-raising highlights the use of language loaded with gendered meaning, it does not explain the implications or significance of the gender-based rhetoric surrounding this environmental group, nor the meaning of women's activism in Pittsburgh. GASP's female members and leaders became civically active and expressed their environmental goals using the language of motherly care for two reasons, functionality and identity. First, maternal rhetoric worked: the women of GASP used language emphasizing the morality and appropriateness of motherly care as a rhetorical strategy to legitimize their involvement, respond to critics, and authorize political action in a region that is best described as socially and culturally traditional. As such, maternal rhetoric used widely accepted standards and mores for a novel purpose. Just as critics similarly emphasized the feminine nature of the group as an attempt to discredit it, the emphasis on maternalism within the group privileged new environmental arguments. Images of school children in surgical masks, expressions of concern for future generations, a mascot who cared for a sick child; these were examples of maternalist rhetoric, emphasizing motherly care for families and dependents. While the term *maternalism* came to have negative connotation for some feminists in the 1970s, it usefully characterizes language used for more than a century by women consciously attempting to gain access to political debates, with the goal of swaying decision making in ways that addressed and benefited women and those in their care.

Maternalism is thus a tool for access rather than an essentialist label, as historians of the Progressive Era have pointed out. Robyn Muncy writes that female reformers seeking to gain access to professions "discovered that their male counterparts were much more willing to cede professional territory, to acknowledge the female rights to expertise in instances where women and children were the only clients." This led female reformers to shape their language, intentionally or not, in ways that highlighted maternal care. As Theda Skocpol argues in *Protecting Soldiers and Mothers*, "Women aimed to extend the domestic morality of the nineteenth century's 'separate sphere' for women into the nation's public life. For a while, this vision was a remarkable source of moral energy and political leverage for the female instigators." This analysis of maternalism emphasizes its function as "political leverage" in gaining access to policy debates, as Seth Koven and Sonya Michel write: "To different degrees . . . maternalism was one of women's chief avenues into the public sphere." Skocpol in fact spells this out later in *Soldiers and Mothers*: "The maternalist *rhetoric* used by politically active women was surprisingly effective with civic leaders and legislators. . . . Symbols of motherhood and domesticity were invoked," with the result that even when advocacy might otherwise be ignored, "women's groups could at least gain a public hearing for their maternalist arguments."[66]

This functionalist use of maternal rhetoric is echoed in histories of civic reform movements and social welfare reform. This concept shows up repeatedly among historians chronicling the various reform movements of the Progressive Era, including temperance, juvenile justice, child labor, and public health.[67] The rhetorical strategy known as *municipal housekeeping* was employed by female civic activists to expand their social role to include political action, expanding the woman's culturally assigned domestic duties to encompass the entire city.[68]

Second, maternal rhetoric appeared because it came from activists who lived in the city as women and mothers. The female activists of GASP shared a largely unspoken vision of a livable, ordered, and just city because they experienced the city in ways different from other civic activists, city planners, or politicians. Their urban reform goals, tactics, and actions emanated from those experiences. The women of GASP spoke about maternal care because they were mothers themselves; they emphasized equal representation for all citizens because they had been excluded from political power themselves; they spoke of care for future generations in a healthy, clean city because that was their shared vision of the future Pittsburgh. This was not a new phenomenon. Even though there was no consciousness of this in the group, GASP's maternalist rhetoric echoed that used by Pittsburgh's

Progressive Era groups, such as the Women's Health Protective Association, Civic Club, and Twentieth Century Club.

In fact, there is a commonly shared vision of urban reform among women both before and after World War II. In her work on women's and men's civic groups in turn-of the century Chicago, historian Maureen A. Flanagan argues that Progressive Era women offered a unique, female vision of "the city livable." In other words, "Chicago women . . . designed a course of urban environmentalism aimed at making the city a livable environment that worked for all its people." Flanagan, Angela Gugliotta, and other historians have argued that women were drawn into the antismoke movement as a result of smoke's disproportionate impact on women's prescribed roles as protector and nurturer of family life.[69]

The experiences of motherhood and women in the city motivated similar women's activism for social welfare, racial justice, and peace in the decades following World War II. In describing the wide variety of women's social activism from the 1940s to the 1960s, Susan Lynn has described a "postwar progressive coalition" of like-minded women. This coalition provided a "bridge that linked the prewar progressive work of women reformers with women's activism in the civil rights, antiwar, and feminist movement of the 1960s."[70] Amy Swerdlow, in describing the postwar organization Women Strike for Peace, argues that a variety of women were drawn to the group because of their shared experiences: "These women rallied to WSP the moment they heard of it, because they were searching for a space in which their moral, political, and maternal stance could be translated into actions" that would not be hampered by existing, male-dominated political organizations. Educated, middle-class women joined the WSP in part because "their devotion to children extended far beyond their own."[71] For both these historians, women's experiences prepared them for activist lives. The experience of the women of the WSP is similar to other women's organizations, such as the League of Women Voters and General Federation of Women's Clubs.[72]

The brief biographies of Michelle Madoff, Marilyn Janocko, Pat Newman, Ann Cardinal, Pat Pelkofer, and Jeannette Widom indicate that, like other Progressive Era and postwar activists, the women of GASP did indeed share a number of traits that prepared them for political action and shaped their experience of the urban environment. They were educated and skilled. They were mothers and civic activists. They were married to professional men, part of an active social life, and possessed considerable social capital. Through this, they became active in city politics and instrumental in the transformation of Pittsburgh. Their activism, in means and method, reflected their lives: it focused upon their experience of the city as mothers and urban dwellers. They found an active and defiant political life on the way to creating a city that reflected their values, a city that valued the health and lives of women and those dependent upon them.

GASP activists used gendered rhetoric because it reflected a divergent vision of the city, a plan for Pittsburgh that could not be evoked with the language of the conservation of resources, efficiency, environmental order, or civic growth. The women of GASP believed that the end result of cleaning urban skies affected women's lives. Their vision of the city's future did not specifically rest upon the continuation of the city's past self-image as industrial powerhouse. Language from the part of GASP organized and staffed almost entirely by women—the educational and organizational sides of the group, as opposed to the scientific and legal committees staffed entirely by men—produced rhetoric detailing an urban vision not present in other GASP statements. While the GASP members attending Variance Board meetings spoke of the specifics of mechanical stokers, coal-fired boilers, slag heaps, and sulfur dioxide measurements, the women of GASP's educational side spoke about children's health, care for future generations, and a maternal mascot weaving through polluted air in search of a clean, livable city with air to breathe.

While the ACCD argued for controlling smoke as a means of boosting the city's self-image and economic viability, GASP's women activists argued for pollution control based upon the health and well-being of women and their dependents. The end results might have been similar, but the means and ideology could not have been more different. Just as the business-oriented members of the ACCD expressed an urban reform vision rooted in their experience as businessmen and civic leaders, the educated, civically active women of GASP expressed an urban reform vision rooted in their experience as mothers and homemakers. In choosing to speak about environmental issues as mothers, the female activists of GASP were explicitly prioritizing the health and well-being of children, their families, and the city's populace over economic development. Their organizational work capitalized on the existing social network of female reform, and the rhetoric of the organization gave voice to a competing, gendered vision of the city and its government in which the group's members declared themselves to be responsible for the environment and those living in it. In short, they proclaimed themselves to be mothers of urban skies.

This panoramic photo of Pittsburgh, and others like it, was used by the Allegheny Conference on Community Development to promote local smoke control in the 1940s and 50s. The Ohio River in the foreground is formed at the confluence of the Allegheny, left, and Monongahela, right. Courtesy Carnegie Library of Pittsburgh.

The Jones & Laughlin Steel Corporation steel mills on both sides of the Monongahela River in Pittsburgh itself, from the Boulevard of the Allies. This image, from November 1947, predates both the smoke control of the 1950s and tougher air pollution control of the 1970s. Courtesy Carnegie Library of Pittsburgh.

The *Pittsburgh Press* caption for this image notes that "A little 'air pollution' is appropriate as commissioners adopt new county code" on December 17, 1969. *Left to right:* Thomas J. Foerster, Leonard C. Staisey, Dr. William R. Hunt. Copyright © *Pittsburgh Post-Gazette*, 2009, all rights reserved. Reprinted with permission.

GASP member Mrs. Irving Nadelhaft selling cans of clean air at the Jenkins Arcade table, January 16, 1970. This picture appeared on the front page of the *Pittsburgh Press* to illustrate the news that a winter inversion had trapped pollution at ground level, exceeding newly set emergency levels and prompting orders to temporarily close polluting industry. Copyright © *Pittsburgh Post-Gazette*, 2009, all rights reserved. Reprinted with permission.

GASP founder and first president Michelle Madoff, never shy about expressing her opinion, makes her feelings clear. Used by permission of the Archives of Industrial Society, Archives Service Center, University of Pittsburgh.

Pat Newman, at the time GASP's publicity chair and later its second president, speaks at the microphone for a hearing on the proposed Article XVIII, May 12, 1972. The *Pittsburgh Post-Gazette* quotes Newman: "Every time a polluter claims he'll have to cut back or close a facility due to pollution control laws, he should be required to open his financial records." Copyright © *Pittsburgh Post-Gazette*, 2009, all rights reserved. Reprinted with permission.

Jeannette Widom signs a fund-raising cookbook at a GASP information table. The cookbooks provided a surprising amount of funding for the group, but more importantly linked GASP to both the corporate world and a dense network of women's civic organizations. Used by permission of the Archives of Industrial Society, Archives Service Center, University of Pittsburgh.

Mrs. Johanna Hicken and two-year-old son, Robert, in the *Pittsburgh Press* on January 22, 1971. The accompanying story says that Hicken believed that individuals should take action: "That's why I'm joining [GASP]. They seem to be doing something effective to clean up our air." Copyright © *Pittsburgh Post-Gazette*, 2009, all rights reserved. Reprinted with permission.

A cartoon image of GASP mascot Dirtie Gertie and brood perched in their nest, located precariously close to industrial sources of pollution. This still image is from one of the animated public service announcements that appeared on local television. Courtesy of the Group Against Smog and Pollution, Inc.

An unidentified GASP member—there is still some evident embarrassment about who, exactly, wore the costume at different occasions—appears as Dirtie Gertie, caring for an off-balance child costumed as Dirty Dick, the Poor Polluted Chick. The costume was a part of GASP's educational mission, emphasizing maternal as well as environmental themes. Courtesy of the Group Against Smog and Pollution, Inc.

Allegheny County Health Department Bureau of Air Pollution Control engineer and GASP member Bernard "Bernie" Bloom, pictured July 21, 1972. Behind Bloom is the Bureau's computer, used to compile air pollution monitoring network data and compile the daily air pollution index, a metric that Bloom created. Copyright © *Pittsburgh Post-Gazette*, 2009, all rights reserved. Reprinted with permission.

The Clairton Coke Works of U.S. Steel, widely described as the largest in the world, spread along the Monongahela River, June 27, 1976. Copyright © *Pittsburgh Post-Gazette*, 2009, all rights reserved. Reprinted with permission.

Smoky Clairton, April 18, 1973. Clairton was a particularly difficult industrial source to regulate; its size, economic importance, U.S. Steel's intransigence, and the profoundly dirty process of coking made the site emblematic of all industrial air pollution in the region, even as other sources were successfully regulated. Copyright © *Pittsburgh Post-Gazette,* 2009, all rights reserved. Reprinted with permission.

EPA official Ed Furia ducks into a car on his whirlwind 1972 tour of the Clairton Coke Works, as reporters look on. Furia's intervention in the Clairton case exasperated all local stakeholders, who closed ranks to defend the locally negotiated consent decree against federal involvement. Copyright © *Pittsburgh Post-Gazette,* 2009, all rights reserved. Reprinted with permission.

In a special session of the consent decree hearings convened at the U.S. Steel Clairton works, Judge Silvestri Silvestri (center) tours the facilities with a team of county inspectors, April 17, 1973. All in the photo are wearing respirators around their necks; behind them, the vertical banks of coke ovens form one end of a long battery, one of twenty at the time. Courtesy *McKeesport Daily News*.

Using language and logic reminiscent of environmental activists, antiregulation groups promote jobs over environment on November 22, 1976. COKE (Clairton Organizes to Keep Employment), accused by GASP of being a false-front organization, is here protesting the 1976 Clairton consent decree's attempt to force U.S. Steel into investing in air pollution control. Copyright © *Pittsburgh Post-Gazette*, 2009, all rights reserved. Reprinted with permission.

5 "Where the Rubber Meets the Road"

IMPLEMENTATION AND THE RHETORIC OF SCIENTIFIC EXPERTISE AT THE VARIANCE BOARD, 1970–1975

It seemed that the real world had intruded upon the remote quiet of the courtroom.

—GASP *Hotline*, 1971

After the excitement and emotion of the first round of public hearings on proposed air pollution control policies in the fall of 1969, the more prosaic process of implementing and enforcing those regulations fell to the county's Variance Board. Yet those board meetings were far from boring. They were extensively reported in the local press and often drew crowds large enough to warrant evening meetings and necessitating the rental of auditoriums normally reserved for musical performances. They were full of fireworks, too, as representatives of the public cross-examined and harried government and industry alike. Citizen environmentalists called for major public utilities to "be padlocked" until they controlled pollution. GASP lawyers "grilled" corporate spokesmen "relentlessly," and subjected officials to "cross examination" in "bitter exchanges." Requests for variance from respected local institutions "drew sharp criticism" and their spokesmen were "chided" for their temerity. Board members confronted corporate lawyers with surprise evidence, challenges to previous statements, slideshows of smoke plumes, and complaints from small polluters who objected to perceived lenience for large polluters. But behind the drama at the Variance Board, an important step in the policy process was occurring. The new county air pollution code had been publicly debated and made more stringent through public outcry. A newly pluralistic and litigious process likewise shaped implementation of the new pollution code, and the result was, most likely, not exactly what anyone had intended.[1]

Historians and political scientists alike expend a great deal of effort examining the creation of policy and the formation of institutions, but little time examining day-to-day enforcement, implementation, or institutional practice. This is unfortunate, since several observers of environmental affairs have noted that "regulations only reveal their true meaning through the enforcement process." One EPA insider has argued that in contrast to policymaking, enforcement is "where the rubber meets the road and everything else hits the fan."[2] In Allegheny County, a close examination of the day-to-day process of enforcement raises a number of questions about the "tough" new air pollution regulations that specifically allowed for citizen involvement. Who were the citizens who actually became involved after the right to intervene was established? Was scientific expertise the most successful rhetoric before the board, or was it trumped by environmental, populist, or economic concerns? And really, what difference did public participation make?

Answering these questions requires avoiding the very real danger of romanticizing public involvement. After all, this is the exact reason that the language of participatory democracy and citizens' rights was so attractive to environmentalists like GASP: who can object to things like *democracy, citizenship*, or the *power of the people*? The reality might be somewhat different than the rhetoric. Even though some government agencies and officials occasionally worked to increase public involvement, for many government officials it is likely that more substantive inclusion might be, as it was for one county air pollution engineer, "a pain in the ear."[3] Involving the public or its representatives in the business meetings of regulators could create pressure for progress, but almost certainly meant more confrontation, more publicity, and the very real possibility of more serious legal complications. But was it worth the pain? Did including citizen environmentalists in implementation and enforcement make a substantial difference in Pittsburgh's pollution? The answer is yes, up to a point.

The Devil in the Details

When county commissioners rewrote portions of the Allegheny County Code in December 1969, they produced regulations creating a specific type of administrative body: an independent board authorized by law and charged with enforcing regulation. Article XVII continued a number of policies from previous versions of air pollution regulation, imposing a variety of restrictions on any producer of smoke or air pollution in the county and limiting the production of visible smoke, measurable particulates, or high levels of sulfur dioxide. The Bureau of Air Pollution Control (or just the Bureau), a division of the Allegheny County Health Department, was charged with investigating and enforcing violations, but neither the Bureau's involvement nor the air pollution limits enumerated in the code were significantly different from the previous incarnations of the Municipal Smoke Control office. However, Article XVII also called for the creation of the independent Air Pollution Appeals and Variance Review Board, an organization that heard requests from polluters for negotiated settlements or lenience in the enforcement of air pollution rules. Colloquially known as the Variance Board, the new entity allowed the Bureau to distribute responsibility for the contentious enforcement process, smoothing the transition to the new code while creating an appearance of reasoned and rational give-and-take and allowing for substantive but controlled inclusion of environmental activism. In short, the Variance Board was an experimental compromise allowing the county to appear firm in pollution control regulations, to remain flexible when it came to enforcement, and yet to be seen as responsive to the public.[4]

To be blunt, industry didn't expect very much from the new board. As an anonymous business leader told *Business Week*, "The board will probably do three or four rash things, then settle down." The idea wasn't entirely new to the region, either: this variance board created for air pollution was similar to the variance boards created to address the environmental impacts of mine acid drainage, a problem that had plagued water quality throughout western Pennsylvania. Still, it was the first time that this mechanism was used to address air pollution in Pittsburgh.[5]

During the 1960s and 1970s the county's health department became the sole governmental agency regulating air pollution on the local level. Federal and state air pollution laws were superseded by local control in Allegheny County (and only in that county) while previous municipal codes enacted by Pittsburgh were abolished in favor of county-level enforcement. The ACHD dedicated larger and larger portions of its total budget to the Bureau of Air Pollution Control over the decade beginning in 1965, and also dedicated proportionally more time in its public reports to accounting for air pollution regulation and enforcement. When county commissioners produced the revised air pollution code known as Article XVII in December 1969, the health department took responsibility not only for monitoring and enforcing stringent pollution requirements but also for encouraging and controlling meaningful public participation in air pollution control. The health department channeled public involvement in the process into the long-standing participation of public representatives on the Air Pollution Control Advisory Board and in the hearings before the Variance Board.[6]

Like its predecessor Article XIII, Article XVII was widely touted as a tough antipollution code. "One of the most stringent in the nation," was a common boast that appeared in the 1969 ACHD annual report as well as in countless journalistic accounts. The new regulations did set strict standards while allowing the health department flexibility in enforcement. For the first time, the new regulations limited sulfur compounds in fuel or emitted smoke, setting a maximum of 1.5 percent sulfur content in the coal burned in steam plants, boilers, or industrial processes. The new code explicitly banned the open-air quenching of slag. This was the common practice of dumping dirty process water on hot material from coking ovens, which produced clouds of smoke and steam containing numerous contaminants.[7] Article XVII stated that "water quenching of slag . . . is prohibited unless the water quenching of slag is performed under conditions which prevent the discharge of all hydrogen sulfide or other air contaminants into the open air."[8]

In addition to these new requirements Article XVII prohibited open trash burning, introduced particulate limits, and allowed the health department power to assess fines (at one hundred dollars a day) and in some cases to appeal for court injunction to shut down polluters. All of these new regulations and expanded powers were lauded in the local and national press as being appropriately strict and comparatively tough on polluters. Still, Pittsburgh-area industrialists knew that the makeup of the board doing the enforcement was more important than the original law; as one anonymous informant said to *Business Week*, "We're not worried too much [about the new clean air code]. It all depends on who is appointed to the board." Knowing that, GASP took special effort to make "sure that the Variance Board established by the new regulations would not be composed of political hacks or friends of the polluters," grooming possible members and lobbying for their preferred nominees.[9]

The strict regulations of Article XVII were substantially transformed when the newly appointed members of the board were introduced into the day-to-day process of air pollution control. In reality, the complex process of variance application and renewal allowed the county opportunities to refashion the law for greater flexibility in dealing with the wide variety of variance applicants. In effect, the Variance Board began to make up its own practices within the confines of the enabling law, inventing multiple time limits, extensions, and reporting procedures.

The Variance Board met every few months throughout the early 1970s in the manner of an informal court of law. The Gold Room, a hearing room on the fourth floor of the Allegheny County Courthouse in downtown Pittsburgh, was designed as a space for large public hearings, and resembled a courtroom in layout, with a raised platform on one side for the five members of the board. Seated behind a judge's bench on the platform were Jean Nickeson, a Pittsburgh citizen and long-time member of the Air Pollution Control Advisory Committee; Daniel Bienstock, an administrator at the U.S. Bureau of Mines research center on the Carnegie-Mellon University campus; chair and Duquesne University law professor Dr. Robert Broughton; Robert Totten, a pathologist at Presbyterian-University Hospital; and consulting engineer Emerson Venable.[10] On the opposite side of the room from the board members sat three separate tables, one for the public (mostly represented by GASP), one for petitioners, and one for representatives of the county's Bureau of Air Pollution Control. Between the two long sides of the room sat the board's secretary (and Bureau air pollution engineer) Carl J. Nim, and a court stenographer who produced full transcripts of all hearings. On the left-hand side of the board members was an enclosed area for journalists.[11] In this courtroom setting, the board conducted hearings, allowing petitioners to present their case along with supporting documentation and testimony but also interrogating witnesses and allowing representatives of GASP and the Bureau to cross-examine witnesses on their own. Symbolically, the representatives of county government sat on the same level as petitioners like U.S. Steel and intervenors like GASP, facing the members

Table 5.1 *Article XVII Variance Petitions, by Category, with Outcomes (1970–December 1971)*

					Resolution		
Category	Petitions	Granted	Denied	Continued	Compliance	None	Special
Industrial							
Iron and steel	25	25	–	–	–	–	–
Coal and coke	6	3	–	–	2	1	–
Slag	8	8	–	–	–	–	–
Chemical	5	5	–	–	–	–	–
Coal-fired boilers	20	17	2	–	1	–	–
Construction	12	6	2	1	3	–	–
Miscellaneous	10	10	–	–	–	–	–
Total (industrial)	86	74	4	1	6	1	–
City and town governments	25	–	20	–	–	1	4
Schools	4	4	–	–	–	–	–
Hospitals	7	6	–	–	1	–	–
Unclassified	8	7	1	–	–	–	–
Total (all)	130	91	25	1	7	2	4
		70%	19%	1%	5%	2%	3%

Note: Analysis of report, "Allegheny County Air Pollution Appeals and Variance Review Board," prepared by Daniel Bienstock, December 23, 1971, ff "Variance Board," GASP Internal.

of the board and surrendering their authority to them—or avoiding the blame, depending upon one's viewpoint.

In its first hearing in the Gold Room, the board's actions presaged many of its later themes: finding innovative solutions to pollution enforcement problems that were not specifically allowed in the text of Article XVII, acknowledging the slow transformation of pollution sources in the county, denying a very small percentage of variance requests, and embarking on a long and litigious relationship with large corporations such as U.S. Steel. Under Article XVII, new regulations for polluters went into effect on January 1, 1970, and individuals, institutions, and corporations began to appeal for variance from the requirements by March. On April 27, the Variance Board heard its first batch of appeals from seven polluters—and promptly began changing the regulations outlined in Article XVII, offering a blanket variance for trash burning in eighty-one separate municipalities. There was no mention of broad, blanket variances with attached conditions for multiple polluters in the text of Article XVII. To the contrary, there *was* a mandate for the immediate halt of open trash burning. Along with the municipal burning variance, which covered three of the first seven petitions, the board denied two petitions (from construction and building supply companies requesting open burning), acknowledged the withdrawal of a petition from the Marquette Cement Company, which had chosen instead to alter its process, and issued its first one-year variance (repeatedly reviewed and extended) to the Universal Atlas Cement Company, a division of U.S. Steel. This first run-in with U.S. Steel prefigured a relationship that threatened to overshadow much of what the board accomplished.

In its first two years, the Variance Board heard 130 requests for relief from the requirements of Article XVII. Those 130 petitions represented a variety of air pollution sources, including the burning of yard waste by homeowners and city governments, medical waste incinerators at hospitals, steam-heat boilers at schools, and a funeral home crematory. But the majority of petitions came from air pollution sources related to industrial production. Table 5.1 shows a tabulation of those hearings by category: each petition hearing number represents a pollution source from a facility owned by a corporation or organization, or a group of pollution sources from a facility that operated as a component of a larger corporation or organization. For example, when Mesta Machine Company petitioned in May 1970 for time to replace its three twenty-five-ton coal-fired air furnaces with a single fifty-ton electric arc furnace with built-in air pollution controls, that petition was heard in a single meeting. But when Jones & Laughlin Steel submitted petitions for seven of its component sites in the county in 1971, each of those received a separate hearing, one each for three boilers (one at the South Side Works, one at the By-Product Coke Department in Hazelwood, and another at the Blast Furnace Department on Second Avenue), an iron cupola, an open hearth furnace, scarfers, and slag quenching at the J&L Hazelwood site.[12]

For a groundbreaking experiment widely touted as enforcing the most stringent air pollution code in the nation, the Variance Board appeared to be quite lenient toward industrial sources of pollution in its initial phase. In the crucial first years of the body's existence it granted 70 percent of all petitions for variance while denying almost no petitions from industrial polluters. For example, of the 130 petitions submitted to the Variance Board under Article XVII between April 1970 and December 1971, 19 percent of them were denied, but less than 5 percent of variance petitions from the many industrial sources of air pollution were denied. The board granted all 25 variance requests from processes directly related to the production of iron and steel, while only 4 variances from industrial pollution sources were rejected outright.

By comparison, the Variance Board was quite ready to be tough when dealing with municipal governments. The bulk of variance denials (20 of 25 total) came in just two hearing days. On January 18 and July 15, 1971, the board turned down variance petitions from twenty different townships. The municipalities had attempted to band together to apply for indefinite variances for open trash fires at town dumps, trash incinerators operated by private waste companies, and open burning of leaves or yard waste by homeowners. All 20 variances were denied en masse, over the protests of the municipalities. The board might have felt secure in issuing this large-scale denial since it had previously issued the sweeping, one-time only variance for open burning in eighty-one separate municipalities. In the opinion of the board, this original variance gave cities and towns eight months to educate residents and organize alternatives to trash burning, and the subsequent variance petitions for municipal trash burning met with resounding defeats—to the outraged cries of the municipalities. This issue became a bit of a political football between the representatives of these small municipalities, who complained of discriminatory treatment compared to industry, and the county government. Pittsburgh mayor Pete Flaherty ended up weighing in on the side of the Variance Board on this matter, to the dismay of county commissioner Tom Foerster, who "tried to heave [him] off the ecology bandwagon" according to the *Post-Gazette*, saying "He's one of those new politics people who are trying to cash in on the environmental issue."[13]

Taken as a whole, the variance petitions listed in Table 5.1 offer a cross-section of pollution sources in Allegheny County. The sample is limited to polluters who were observed by enforcement officers to exceed county code and who then applied for a variance to continue operations. Even so, the variance petitions submitted under Article XVII included the many possible types of air pollution sources in Pittsburgh and its environs. From Abrasives Metal Company to Woodville State Hospital, the variance requests cover a wide variety of industrial, commercial, and residential pollution sources. While most variance requests were from industrial concerns related to the production of iron and steel, the variety of pollution sources is revealed in the list of variance requests, some of which might not have been previously apparent to pollution regulators. For example, the board discovered a previously unforeseen type of air pollution source when it determined that the Robert M. Chambers Company of Monroeville contributed to air pollution with its use of open wood fires, built underneath railroad cars filled with coke to thaw the frozen contents before dumping. Once the company switched to coke-fired "salamanders" (or small metal stoves used to heat outdoor construction) the board determined that no variance was necessary.[14]

Carnegie-Mellon University was one of several schools that came before the Variance Board under Article XVII; even more schools with integrated coal-fired steam-heating plants filed variance petitions under the subsequent Article XVIII when it became clear that the health department intended that all city and county government facilities should comply with the new regulations. Carnegie-Mellon received a one-year variance under Article XVII, and updated its coal-fired steam heating equipment while attempting to purchase low-sulfur coal. But when the school returned to ask for another variance in 1973 under Article XVIII, it received a tongue-lashing instead: Michelle Madoff declared the school's attitude "immoral," and rapped its knuckles: "you should be a leader and be setting an example to industry and the schools instead of being a contributing factor to air pollution," she lectured the university's representative. Board member Emerson Venable declared that since all county households had met emission standards, it would be unjust to allow Carnegie-Mellon to slide.[15]

The board often divided variance requests into separate docket numbers applying to different pollution sources or sites. Table 5.2 shows a summary of individual dockets in which variances were issued, and indicates that the board appeared to offer petitioners multiple opportunities to come into compliance, granting variances to most who applied and extending variance periods for 49 percent of all those who originally received a variance. The board exercised wide leeway to require progress reports and impose additional conditions on granted variances: both of these practices appear to have become commonplace throughout 1970 and 1971 even though they were not specifically enumerated as rights of the Variance Board under Article XVII.

Variance Board members exercised flexibility in both the type of variance permits they awarded and the time allowed to come into compliance. They even developed a shorter variance period—three months—than was allowed for in Article XVII. In the words of one board member: "You know, you come in and you want a year to make plans. Basically, you can have ninety days to make plans. I mean, ninety days to make a decision—to go out of business, or if you're going to control it [pollution], or what you're going to do. That's kind of a transitional category."[16] The three-month variance came to "constitute the bulk of the decisions," according to members of the board. These short, three-month variances

Table 5.2 *Article XVII Variances Granted, by Docket, With Conditions*

	Dockets Receiving Variance	Extensions	Conditions	Progress Reports
Industrial				
Iron and steel	30	13	14	19
Coal and coke	5	3	3	4
Slag	7	7	7	7
Chemical	5	3	4	5
Coal-fired boilers	23	8	8	22
Construction	9	6	8	–
Miscellaneous	12	7	7	12
Total (industrial)	91	47	51	69
City and town governments	–	–	–	–
Schools	4	2	3	4
Hospitals	6	1	2	5
Unclassified	7	3	3	6
Total (all)	108	53	59	84

Note: Data compiled from "Allegheny County Air Pollution Appeals and Variance Review Board," prepared by Daniel Bienstock, December 23, 1971, in ff "Variance Board," GASP Internal; information for individual docket number (where multiple dockets are issued for single petitioner) from "Variances: Conditions—Results—Compliance," compiled by Carl J. Nim, Jr., ACHD air pollution engineer, November 1971, in ff "Variance Board," GASP Internal.

required active oversight from the board. Though almost all petitioners requested the maximum time period legally allowed (one year; some even requested two or three years, far beyond what county code allowed), 45 of the 108 variances granted in individual dockets were for three-month periods, bringing down the median length of variances granted to six months or less in all industrial categories, as depicted in Table 5.3. Put another way, the median time period granted was half of what was requested. Practically, however, in the 45 cases where three-month variances were granted, 26 were extended as least once, 28 included conditions beyond those enumerated in Article XVII, and 37 required continued progress reports to the board. None of these practices were outlined in the county regulations, but these shorter variances gave the board leverage to expand its oversight power and involved the group as an industrial watchdog, calling to check with the company to see if it had installed new equipment or improved fuel quality. As a board member noted: "We then must follow up to see what has been done. Much of our activity is that of telephoning to see what has been done."[17]

This sort of creative, flexible, and intensive oversight on the part of the Variance Board was repeatedly challenged by multiple parties. For entirely different reasons,

Table 5.3 *Article XVII, Median Length of Variances in Months*

	Requested	Granted
Industrial		
Iron and steel	12	6
Coal and coke	12	6
Slag	6	6
Chemical	12	9
Coal-fired boilers	12	6
Construction	12	3
Miscellaneous	12	3
Median (industrial)	12	6
City and town governments	–	–
Schools	12	12
Hospitals	7	12
Unclassified	5	12
Median (all)	12	6

Note: See sources for Table 5.2.

both the commonwealth of Pennsylvania and U.S. Steel objected to a variance granted to the steelmaker's Edgar Thomson works in the fall of 1970. U.S. Steel had argued against the original conditions of the variance when the decision was made in September, asserting that "there is no enabling legislation supporting the procedure followed by the board." By February of the next year, U.S. Steel was objecting to the requirements attached to the variance, arguing in an appeal to the Common Pleas Court that not only was there no authority in Article XVII for such oversight but that the Variance Board itself represented an illegal delegation of government power to an independent body. From the opposite side, Assistant Attorney General Marvin Fein argued that no variance should have been granted in the first place, and that U.S. Steel was abusing the process "to continue to spew out an indeterminate amount of pollutants completely unmolested." He continued: "It is very difficult to understand how U.S. Steel could have the audacity to file an appeal here after having obtained all of the advantages of a variance from Jan. 1, 1970 to the present." The county solicitor, for his part, was forced to defend both the variance decision and the Variance Board itself to the Common Pleas court judge.[18]

There are clear limits to analyzing the actions of the Variance Board through a compilation of hearing statistics. While the findings above seem to indicate flexibility on the part of the Variance Board and serious improvements by petitioners to meet county code, the most politically contentious and legally fractious case does not even appear in tabulations of variances granted or variances denied.

Permit hearing 52, docket numbers 112 and 143, were related to U.S. Steel Corporation's use of quench water in its coking process at Clairton, a practice specifically prohibited by Article XVII. Docket 112 was settled with the addition of baffles on several quenching towers to contain vapors and steam from contaminated quench water. But records for docket number 143 (which was discussed on August 5, 1970, November 16, 1970, and June 9, 1971) show no variance approval, denial, timetable, or conditions. In fact, U.S. Steel's petition to use contaminated quench water in its coking process became the most public and divisive of all the cases before the Variance Board, but in simple analysis of board decisions it is largely invisible. While Article XVII specifically prohibited the open release of contaminants from water slag quenching, U.S. Steel continued the practice throughout the early 1970s. The board made no decision on the docket through 1970 and most of 1971, though it heard from technical, scientific, and industrial advisors on the matter. But the long delay meant a de facto two-year variance for U.S. Steel even before the matter worked its way through the courts for an additional three years. Through this entire period U.S. Steel and other steelmakers in the Pittsburgh area continued the ostensibly illegal practice of open-air water quenching. This issue appears as the single unresolved item in the "Industrial" column of Table 5.1. This was the devil hidden in the details, and it led to considerable difficulty for all parties in the future.

Actors on Stage

From a historical perspective, the institution of the Variance Board represents the creation of a novel public space in the political landscape. Less structured than the courts, somewhat independent from county government, separate from political parties, more public than the "gentleman's agreement" of previous decades, and yet less ephemeral than the debates in local media outlets, the Variance Board allowed for a well-ordered argument on environmental, economic, and public health issues. The board was a nexus of three competing interests: the public, the industrial corporations, and the county public health bureaucracy. These three interests interacted on the stage of the Variance Board before several audiences, including the board members themselves, the general public, and the representatives of media who interpreted and transmitted the actions of the board to the greater public.[19]

On the public stage provided by the Variance Board, the three competing interests each inhabited roles, each with characteristics largely predetermined by the board itself. Corporate and industrial interests appeared before the board as petitioners or supplicants, and adverse decisions were often made by the board based upon the attitudes and responsiveness of industrial representatives as much as on the specifics of the case. Representatives of the public interest (most often GASP leaders) appeared as prosecutors in petition hearings, taking an active role in cross-examining petitioners. Health department officials appeared mostly in an advisory capacity, allowing the public representatives to take the adversarial position and offering evidence or interpretation of regulation only when asked. And the members of the Variance Board itself acted as a tribunal of judges, allowing competition between the prosecution and defense to play out in front of them in hopes of determining the truth.[20]

GASP itself thought of the Variance Board as a public stage. GASP described its representatives as actors filling theatrical roles, noting that a hearing concerning U.S. Steel "played to a packed house," and telling members, "Don't miss the second performance!" A story written by a GASP member in the Pittsburgh *Forum* declared that "the curtain fell on Act I of GASP's dramatic fight against air pollution" with the celebration of the group's first birthday, but that "Act II commences with a very short intermission, Arthur B. [Gorr] told the standing-room-only-audience," as "one of the major antagonists in the drama, Duquesne Light Company, will appear before the [Variance Board]." Another mailing provided a seating chart for the room in which the board met, declaring that "You Can't Tell the Players Without A Program," and that GASP would be "Presenting Act IV of the Clairton Coke Works Quenching Hearing" to an audience made up of concerned citizens. Yet another mailing announced that the next hearing would be at a different time, and that, subsequently, "'the best matinee in town' should be even better in the evening."[21]

By combining the formal structure of a courtroom hearing with the public space of citizen protest, the Variance Board presented a blended space between state and society, something that GASP members observed in their own accounts:

> *Area residents came and stayed as long as they could. Many mothers with small children attended the hearing, along with long-time residents of Clairton and surrounding communities. It seemed that the real world had intruded upon the remote quiet of the courtroom. Polluter representatives appeared somewhat unnerved by the number of concerned citizens they had brought to the hearings. The tension of close questioning was interrupted periodically by the cry of a child. A vocal and articulate representative of Steelworkers Union Local 1557 (Clairton) expressed his disgust with USS's nit-picking examination by snorting out loud.*[22]

The Variance Board was a physical space for the metaphorical interaction of state and society, a local venue for debates in the public sphere. The public sphere itself can be defined as the arena in which citizens have historically exerted their influence over collective decisions. Social critic and philosopher Jurgen Habermas has argued that the public sphere provides a space that can "subject persons or affairs to public reason," and can "make political decisions subject to appeal before the court of public opinion." In this sense the public sphere is defined as open space between the mechanisms of the state and the representatives of society. This description fits the situation in Pittsburgh quite well; while the Variance Board was a creation of county government (i.e., the local state), it was, in intent and function, an independent, quasi-state organization. The board itself was not the same thing as the local state, but rather a collection of experts and citizens with delegated state authority; county commissioners thus attempted to distance themselves from possible backlash from enforcing environmental law on local employers. The board allowed for the substantive inclusion of civil society in the mechanisms of state regulation and enforcement, and did so in a manner that allowed for maximum publicity and visibility. As such it was an ideal space for newly organized citizen environmentalists, empowered by a national sense of the value of grassroots democracy as well as the federal requirements for citizen involvement.[23]

While the Variance Board settled into its role of industrial oversight, public representatives likewise adapted to their roles as prosecutors before the board. This role is most clearly demonstrated by GASP's involvement in the enforcement of air pollution regulations for utilities such as local electricity producer Duquesne Light. Perhaps because Duquesne Light had been the most vocal in protesting the inclusion of public representatives at Variance Board hearings, or possibly because of the importance of the utility's actions to regional air quality, GASP members paid special attention to that company's variance petitions. The result is a surprising record of public involvement in the day-to-day enforcement of air pollution regulation; one that breaks with the county's long history of excluding public representatives from air pollution policymaking.

After the direct challenges to GASP's intervention in the early Variance Board hearing in 1970 had been settled, GASP and Duquesne Light settled into a constant give-and-take before the board. Issues at hand included the endless battle over forcing regulated firms to seek out low-sulfur coal, closing the Colfax Power Station and opening another in Cheswick, and regulating downtown steam-heating plants and other facilities. But even as Duquesne Light and regulators engaged in constant discussion, they were judged by the public to at least be honorably engaged in the process, as compared to their earlier intransigence: "Duquesne Light had a change of heart from 'fight' to 'cooperate' in 1970 after one of its top vice presidents became regularly involved in negotiations with county air quality officials," recalled a *Post-Gazette* editor in 1974.[24]

One particular meeting of public, regulator, and regulated usefully illustrates some of the subtleties and strangeness of this interaction. At 9:00 in the morning on January 24, 1972, the Variance Board met, as it often did, in the Gold Room of the county courthouse in downtown Pittsburgh. This time, the board met to discuss pollution from Duquesne Light's James Reed Power Plant. The Reed power plant was located directly west of the city, on Brunot Island. The island itself was in the middle of the Ohio River, thus making the plant prominently visible from downtown. The hearing was a continuation of a variance petition submitted on July 20, 1970, and granted until 1972 on the condition that the company submit progress reports on planned construction of low-sulfur oil-burning boilers to replace the coal-fired boilers originally used to turn the electricity-producing turbines.[25] GASP appears as the sole intervenor in the hearing transcript, and GASP representatives intensely questioned Duquesne Light. Representatives from the Bureau, though present in the room, allowed GASP to ask thirty-eight out of forty total questions. The transcript is worth reproducing at length, as it demonstrates both the roles inhabited by the participants, the powers of oversight wielded by the board, the adversarial nature of the discourse, the technical details under discussion, and the conspicuous absence of any representative of the public save GASP.

> THE CHAIRMAN [ROBERT BROUGHTON]: *Today's hearing was convened after it was learned by letter from Mr. Lee Love that the plans of Duquesne Light Company with respect to Reed Power Station had changed to such an extent the Board felt it should hear testimony and consider, you know, the question of the continuance of the variance anew.*
>
> *Let's see, GASP, which had intervened in the original petition, is here, and the Bureau is here, so we have some testimony from Duquesne Light, and why not proceed, Mr. Olds?*
>
> MR. [DAVID M.] OLDS [DUQUESNE LIGHT LAWYER]: *If the Board please, I should also note in accordance with prior rulings of the Board we of course did give notice to the other intervenors, also. At this time—*
>
> THE CHAIRMAN: *I suppose I should note at this time there are no other intervenors present; in particular, the City of Pittsburgh.*[26]

GASP members Michelle Madoff, Patricia Newman, and Arthur Gorr proceeded to ask a daunting series of questions, despite being the only public representatives present. The one-hundred-page transcript from that morning's three-hour session features Madoff interrogating both Duquesne Light representatives and Bureau engineers. Here she cross-examines ACHD engineer Joseph J. Chirico, attempting to demonstrate that the nearby Reed Power plant was degrading air quality levels in the downtown area:

> MRS. MADOFF: *We have an index that is given out on a daily basis of the day before, and two days ago the index was 61.*
>
> MR. CHIRICO: *Right.*
>
> MRS. MADOFF: *The downtown station was what?*
>
> MR. CHIRICO: *That we would have to look up. I couldn't tell you that.*
>
> MRS. MADOFF: *I'll tell you. It's over 100.*
>
> MR. CHIRICO: *It's possible, yes.*

MRS. MADOFF: *It's not possible, it was. It was over 100.*

MR. CHIRICO: *All right.*

MRS. MADOFF: *What I'm saying, doesn't Reed contribute to this downtown reading essentially?*

Chirico does not give the direct answer that Madoff wants here, and they continue their exchange for several more pages of the transcript, as Madoff attempts to get the representative of the health department to provide the testimony that GASP wants, namely that the Reed power station is the largest contributing factor to downtown air pollution:

MRS. MADOFF: *It is a factor.*

MR. CHIRICO: *It can be, certainly.*

MRS. MADOFF: *It's right downtown.*

MR. CHIRICO: *No, Reed is not downtown.*

MRS. MADOFF: *It's darn close.*

MR. CHIRICO: *It's three miles away, really. So when you're talking effects, you have to be pretty specific as to under what conditions. . . . Let's just say it's one of the major sources in a three-mile area, or radius.*

MRS. MADOFF: *I won't, no. Who does he work for?*

MR. CHIRICO: *I work for one department. I don't keep changing.*

MRS. MADOFF: *No comment.*[27]

When the Bureau representative didn't provide the direct answer that Madoff was looking for, she and the other GASP members continued to question the Duquesne Light representatives concerning the availability of low-sulfur coal, coal-washing technology to provide low-sulfur fuel, the possibility that Duquesne Light would idle or dismantle the plant in question, and the fine points of combustion turbine technology. From beginning to end, this excerpt demonstrates a number of interesting aspects. Broughton notes that the board felt that it could, on its own, compel Duquesne Light to appear before the board due to new information, demonstrating the oversight power wielded by the board. Once before the board, Duquesne Light was harried and badgered not by county regulators but by GASP, appearing as public intervenors. The county itself was also subject to Madoff's withering sarcasm.

This type of vigorous public representation appears in other Variance Board hearings. The *Post-Gazette* made it sound as if hearings were inquisition by torture for industry, noting in 1970 which companies were next "on the griddle." The *Press* reported a hearing in 1972, noting that local electricity producer West Penn Power had come under fire for continued operation of coal-fired boilers in one section of its facility, none of which had been fitted with pollution controls but all of which the company argued it needed to produce power at high loads. The story continued:

> *Under cross examination, Carl Cater, manager of generating facilities planning for Allegheny Power System, the parent company, admitted section six could be shut down.*
>
> *Attorney Arthur Gorr, counsel for the Group Against Smog and Pollution (GASP), asked Cater: "If section six failed tomorrow what would you do to replace the power it supplies?"*
>
> *Cater replied: "We could do one of two things. We could do nothing and hope to get through that period if we had real good luck. What we probably would try to do would be to purchase power from a neighboring utility."*

Gorr is highlighted in the article as a public representative, "cross-examining" witnesses from industrial concerns, forcing them to admit that it was economically unnecessary for the utilities to operate pollution sources, and demonstrating the availability of pollution control technology. The story goes on to chronicle Gorr's courtroom victory: "Robert A. McDonald, Allegheny Power System president, admitted under questioning by Gorr that dust-collecting equipment was available in 1958 but not installed on the offending stacks." The newspaper story does not indicate that representatives from the Bureau asked any questions of the industrial representatives.[28]

In using the term "cross-examination," identifying Gorr as "counsel," and describing the reluctant industrial representative as "admitting under questioning" facts adverse to their position, the *Press* characterized Gorr as a public prosecutor. Michelle Madoff certainly seemed to take that role in her questioning of the ACHD and Duquesne Light. GASP representatives appeared before the Variance Board not only as defenders of the public interest but also as symbolic public prosecutors, and the journalistic coverage of the events describes GASP in those legalistic, confrontational terms. This type of conflict did not necessarily endear GASP to other participants. One noted, "sometimes they've been more abrasive than I would have wished. You can achieve your end in a gentlemanly or ladylike way, and maybe this wasn't always the case [with GASP]."[29]

Even while it continued active participation in the county air pollution control process, GASP spent an enormous amount of time throughout 1971 and 1972 lobbying for increased public input in the federally required state implementation plan for the 1970 Clean Air Act.[30] Despite this expansion of the group's scope, GASP still was able to organize a great deal of public input for a moderate revision of county air pollution code in the first half of 1972, which was instigated by county officials attempting to come into compliance with newly promulgated Clean Air

Table 5.4 *Article XVIII Hearings in Which GASP Intervened, 1972–1975*

Industrial	
Iron & steel	21
Coal & coke	3
Slag	2
Chemical	2
Coal-fired boilers	8
Construction	10
Petroleum	12
Prepared foods	6
Miscellaneous	14
Total (Industrial)	78
City and town governments	4
Schools	6
Hospitals	4
Total (all)	92

Note: AIS GASP. Categories assigned by author; "Iron and steel" includes all corporate entities primarily producing metals, "coal-fired boilers" includes all electrical utilities and private steam-heating plants, "Schools" and "Hospitals" mostly refer to those institutions that operated their own steam-heating plants.

Act amendments. Following a (by-now) well-traveled path for preparing public testimony, the GASP leadership organized citizen workshops, provided clerical assistance for preparing and submitting the text of public testimony, and facilitated two days of citizen testimony before the county commissioners and the board of health. In general, GASP supported the specific pollution limits suggested by the county, while concentrating on three points designed to further increase public involvement in the process: "1) more citizen participation, 2) better availability to the public of information in the Bureau of Air Pollution Control files, and 3) more adequate reaction from the Bureau to citizen's complaints."[31]

GASP's involvement in the day-to-day process of air pollution control remained largely unchanged once the new Article XVIII became law on June 15, 1972. GASP continued to convey citizen complaints to the Bureau, to sit on the advisory committee, and to provide expert and citizen commentary as intervenors before the Variance Board. There appears to have been no major change in the type of polluters seeking variances during this period, although the record is slightly more fragmentary. However, GASP files for the following three years do indicate that GASP intervened in 92 of 257 total hearings during that time alone, as demonstrated in Table 5.4.

Much like the list of the many petitioners under Article XVII, the list of applications to the Variance Board under Article XVIII covers a wide range of possible pollution sources. Steel, coal, and coke concerns are fully represented: Jones & Laughlin Steel, United States Steel, Shenango Coke, Vulcan Materials, Bethlehem Steel, and Alcoa all petitioned for more time to meet county emission standards. Most of Duquesne Light's major installations had a hearing, as did local storage facilities for national petrochemical companies including Exxon, Gulf, Mobil, Sun, and Texaco. Pittsburgh's long-established food-processing industry also requested extensions; the Armour meatpacking plant and the H.J. Heinz facility both were found to be in violation of county law. However, the Variance Board also heard (and in some cases, denied) requests for variance from large steam-heating plants at schools and apartment complexes. For example, the Allegheny County Housing Authority, Allegheny County Steam Heating, Carnegie-Mellon University, Dixmont State Hospital, Elizabeth Forward School District, Pittsburgh School District, and St. John's Hospital were all found to be in violation of county emission requirements. The delicate balance between large and small polluters, and jobs and environment, continued. The small Louis Hahn and Son nursery was denied a variance to operate coal-fired heaters in its greenhouse. The board's decision, signed by its second chair, Anthony P. Picadio, noted the "Board is appreciative and has always taken cognizance of the financial problems that can beset a small business enterprise, particularly one of marginal subsistence, in installing pollution control equipment. However, in the absence of a firm plan of compliance the Board cannot grant a variance."[32]

It was with these small and medium-size industries, schools, and municipalities that the county had, arguably, its greatest success. In the words of ACHD engineer and GASP member Bernard "Bernie" Bloom, "there were lots of businesses—they would be the largest pollution source in another jurisdiction. . . . [T]here were glass factories, foundries, the Heinz plant near downtown Pittsburgh, lots of them who were mid-size, Duquesne Brewery, Iron City Brewery, had pretty big commercial, industrial boilers; many, many places like that, and the Variance Board was able to noodge them along the way." Even while the press and the public paid more attention to the large industrial employers and polluters who appeared before the Variance Board, the Bureau of Air Pollution Control's engineers made exceptional progress in regulating the steam boilers and coal-fired heating plants scattered throughout the county. "Enforcement people relish penalties—fines, but the agency, led by [Bureau chief Ronald] Chleboski, wanted compliance," says Bloom now, and even if the smaller polluters weren't fined or eliminated, "they were switched to cleaner fuels, and we had reductions."[33]

During the early 1970s the Pittsburgh region witnessed a contentious and occasionally overexposed Variance Board that nonetheless had success in regulating air pollution. Functionally the board allowed for flexible but insistent enforcement.

Politically it removed the responsibility of regulating employers from the hands of the county commissioners, while keeping news of environmental improvement before the public and allowing for structured public involvement. It wasn't pretty, but it was progress.

The Gendered Language of Expertise and Professionalism

There was something different about GASP's interactions in the structured public sphere of the Variance Board. While GASP's female activists were visible at the hearings, and while they were a novel addition to the policymaking process, their presence was not marked by the same rhetoric of maternalism found elsewhere in the group's activism. The language of citizenship, so prominent in arguments over the group's inclusion in policymaking, was also missing from the implementation and enforcement of policy before the Variance Board. Rather than citizenship and maternalism, the Variance Board was the venue of professionalism and expertise, and GASP members brought their credentials and experience to detailed discussions of technical regulation and monitoring. Here, the gender division between GASP's various functions again became obvious. While the group's leadership, membership, and educational wings were largely the province of female activists, the group's medical, scientific, and legal committees were almost exclusively male, and in interactions between regulators and these committees of the group, the language of maternalism was replaced by claims of scientific and technical expertise. Still, the purpose was the same: GASP's activists were laying claim to the right to intervene, whether they spoke of the rights of the citizen, the health of future generations, or claims concerning parts per million and industrial processes.

In many cases before the Variance Board, GASP organized a noteworthy community of local experts to testify. This was not always a seamless process; experts often disagreed on interpretations of the law, judging the best available technology to meet the law's requirements, or choosing pollution indices and measuring equipment. In the case of Shenango Coal and Coke Company's coke ovens, board members argued with GASP-supplied experts. After Carnegie-Mellon University professor of metallurgy and county advisory committee member Robert W. Dunlap appeared before the Variance Board to discuss his paper entitled "The Desulferization of Coke Oven Gas: Technology, Economics, and Regulatory Activity," board member and nationally known chemical engineer Emerson Venable questioned the report, saying it "contains errors which seriously affect the value of the report as a basis for the regulations of the county." The subsequent debate among Dunlap, Venable, and Bureau chief Chleboski resulted in a variance for Shenango because the Variance Board found it difficult to determine whether Shenango was actually violating Article XVIII. Because of the disagreement among the experts, the Variance Board failed even to determine whether Shenango was installing the newest and most efficient control technology.[34]

There were other battles of competing claims. County engineer Bernie Bloom and meteorologist Joe DeNardo engaged in a war of the experts when the local television weatherman, regularly employed as a consultant by government and industry alike, offered contradictory viewpoints before the board concerning a Duquesne Light variance request. When the fact that DeNardo was employed by both the regulator and regulated was downplayed by board member Emerson Venable, "Bloom said later that he considers the matter potentially serious," according to journalistic coverage.[35]

GASP's involvement in the Variance Board hearings rarely caused such confusion, and by 1974 most members of the Variance Board supported GASP's involvement, particularly its offers of technical assistance. One member noted, "[t]hey have helped us by submitting technical reports on specific aspects of pollution. For example, they supplied a report on coke-quenching; on the Edgar Thompson [*sic*] open-hearth use of oxygen lancing they supplied a report which is contradictory to what USS claims." GASP's authorship of expert technical reports had an even greater impact in the case of the U.S. Steel Clairton works, when the organization released an immensely detailed "Task Force Report" on air pollution, economic models, the coking process, and control technology. The impact of the Clairton report inspired Madoff and others to continue this approach, envisioning a "Dirty Dozen" series of technical reports. Under the auspices of the newly organized "Citizens" Environmental Task Force (later Enviro-SOS), an impressive list of independent scientists and experts compiled a catalog of twelve technical reports, each intended to specify the state of the art in available control technology and processes. These reports provided technical details and arguments ready-made to counter industry's protestations that pollution could not be controlled due to lack of appropriate technology. The reports were written by the many academic and professional men associated with both GASP and Enviro-SOS, although authors' names did not appear on the reports. Instead, they were stamped only with the Enviro-SOS letterhead, a mechanism that allowed experts to contribute information without fear for their professional positions, according to Madoff. From "Operation and Maintenance of Coke Oven Doors" to "Control of Sulfur Dioxide Emissions from Coal-Burning Industrial Boilers," to "Sulfur Dioxide Removal by Scrubbing from Power Plants," the reports were extremely technical, citing industry publications, scientific journals, and air pollution standards from across the nation. Enviro-SOS envisioned this as a self-supporting outreach program, and planned to sell the reports to libraries, corporations, agencies, and environmentalists nationwide.[36]

GASP's involvement before the variance board, its ability to make credible statements to the press, and its authorship of reliable technical reports all stemmed from the group's social capital. Its membership was drawn from neighborhoods and institutions that hosted what was, at that moment, the world's preeminent concentration of air pollution experts. This may make the experience of GASP and Pittsburgh unique; after all, members of the public were likely more able to intelligently discuss the complexities of control technology there than anywhere else. But it's more likely that the depth of the talent pool simply matched the seriousness of the problem in Pittsburgh, and that pollution debates in other cities could likewise draw upon local professional expertise.

Regardless of comparative value, Pittsburgh's air pollution debate was based upon a remarkably diffuse population of experts. Independent industrial labs, academic research centers, and a variety of institutions and individuals committed to air pollution monitoring clustered in postwar Pittsburgh, most likely because the region's atmosphere furnished them ample research topics while Pittsburgh's industry furnished them ample employment. The list of these institutions is impressive. The national offices of the Air Pollution Control Association were moved to the Mellon Institute for Industrial Research in the fall of 1951. The Mellon Institute, an independent research laboratory before being folded into the Carnegie Institute of Technology to form Carnegie-Mellon University, was located in the same neoclassical building as the American Iron and Steel Institute and the Industrial Hygiene Foundation.[37] The Instrument Society of America, a national professional organization for environmental engineers, had its offices not far down the street. The International Association for the Prevention of Smoke, a professional organization of engineers, public health experts, and municipal administrators, was founded in Pittsburgh during the Progressive Era, and its successor, the Air and Waste Management Association, was and is headquartered in Pittsburgh.[38] While these were important institutions, the most significant inhabitant of that same Oakland neighborhood was the University of Pittsburgh. Pitt's Graduate School of Public Health was a leading institution in the epidemiology of air pollution, and the institutional home of nationally known air pollution researcher Morton Corn.[39] Just a few blocks east, at what was known as Carnegie-Mellon University after 1967, Lester B. Lave and Eugene Seskin turned out a long series of nationally recognized articles on air pollution and human health.[40] Elsewhere on campus, the Pittsburgh Coal Research Center was funded by the Bureau of Mines, U.S. Department of the Interior, and was the institutional host of Variance Board member Daniel Bienstock. Individuals from these institutions collaborated in the creation of marketable air pollution instrumentation with several Pittsburgh companies: the Research Appliance Company built what was marketed as the AISI air pollution monitor, a reel-to-reel paper tape sampler developed at the American Iron and Steel Institute that became the industry standard throughout the 1950s and 1960s.[41] Across town, the Mine Safety Appliances Corporation adapted its high-volume forced-air samplers from mining for use as ambient air quality sensors.[42] The Westinghouse Electric Corporation had a complex of research parks in the area, outside of Pittsburgh proper to the east, which rarely focused specifically on air pollution matters but still employed a large number of engineers, metallurgists, chemists, and physicists who lived in Pittsburgh and the region. Finally, researchers from these institutions formed several private consulting firms: engineer and Variance Board member Emerson Venable formed Emerson Venable and Associates, and Industrial Hygiene Foundation director of engineering Wesley C. L. Hemeon formed Hemeon Associates to offer consulting services and construct filter paper samplers.[43]

Many of GASP's members were associated with these diverse professional and scientific institutions. Nearly a third of the group's leadership were men listing professional credentials, scientific and otherwise, judging by compilations of group letterhead. Advanced degrees or titles included Ph.D., M.D., J.D., O.D., LL.B., M.S., Esq., and Reverend. Conversely, no female members listed advanced degrees on the letterhead, even though at least one—Esther Kitzes—had a master's degree. Of the ninety-six leadership positions occupied by men with advanced degrees over the five-year period from 1970 to 1975, thirty-nine were occupied by men with Ph.D.s, many of whom worked in technical fields. Most of these were faculty members at Carnegie-Mellon University, the University of Pittsburgh (both in Oakland, the neighborhood directly bordering Squirrel Hill to the west), or Duquesne University, a Catholic school located near downtown Pittsburgh. Lester B. Lave, mentioned above as a nationally known expert on air pollution, was listed as a member of the GASP board in every letterhead produced between 1970 and 1975. Chair of the Economics Department at Carnegie-Mellon and a fellow at the Brookings Institution, he was an active member of the GASP board.[44] Emmanuel "Manny" Sillman was a professor of botany at Duquesne University who had completed his Ph.D. in zoology at the University of Michigan in 1954. He was a regular in GASP's speakers bureau whose rhetorical skills were displayed in the group's first movie, and the chair of the group's scientific committee.[45] Robert Freedman, a graduate of the Massachusetts Institute of Technology, was a research chemist employed at the U.S. Bureau of Mines, with laboratory space on the Carnegie-Mellon campus. Freedman, a founding member of GASP and an appointed member of the county Air Pollution Control Advisory Committee, accompanied Michelle Madoff to work with the spin-off group CETF, later Enviro-SOS. He was also a prominent expert on utility regulation, testifying both locally and to the EPA.[46] Walter Goldburg, an associate professor at the University of Pittsburgh, had a Ph.D. in physics and was an active member both of GASP and the local chapter of the Federation of American Scientists. He served as president of GASP during the late 1970s and again in the year 2000.[47] Dr. Arnold S. Kitzes, married to GASP member Esther Kitzes, was active both on the board and variously as treasurer, vice president, and president during the first five years of GASP. He was a project leader at the Westinghouse Corporation's Test Engineering Section at a site just east of Pittsburgh, coming to the position after ten years at the Oak Ridge National Laboratory.[48] Morton Corn, an attendee at GASP's very first organizational meeting and a contributor for years thereafter, was a nationally recognized and widely published academic focused on occupational health and chemical engineering at the Graduate School of Public Health at the University of Pittsburgh. Like Madoff and Freedman, Corn served for a time as an appointed member of the county Advisory Committee. Corn would go on to be a surprise nominee for federal assistant labor secretary for occupational safety and health, and led the Occupational Safety and Health Administration (OSHA) from 1975 to 1977.[49] There were many others at Pitt's Graduate School of Public Health who lent their expertise to the public air pollution debate, including John Frohliger. He was an active member of GASP and wrote a laudatory article about GASP for a local periodical, and he was also an accomplished researcher and scholar with extensive publications in air pollution chemistry.[50]

In addition to the many doctorates, individuals in the GASP leadership were equally likely to hold an advanced degree in law or medicine, with eleven out of ninety-six listing themselves as "J.D.," "LL.B.," or "Esq." and another eleven listing "M.D." after their names. For example, Arthur R. Gorr was born in Pittsburgh, and after graduating from Princeton and receiving his law degree from the University of Pennsylvania in 1959, he returned to Pittsburgh in 1960 as a junior partner at the law firm of Stein & Winters. Gorr was a longtime board member of GASP, served as vice president from June 1971 to June 1972, and was president in 1974 and 1975. There was also considerable expertise when it came to the medical profession; Dr. Charles G. Watson, the head of GASP's medical committee, was an active member of the Allegheny County Medical Society, served as the chair of that organization's Committee for Environmental Problems, and often submitted GASP articles or announcements to that organization's newsletters.[51]

Arguably, the most important of those GASP members with law degrees was Robert Broughton, the very same Harvard Law graduate, Duquesne University law professor, and Variance Board chair who had made the decision to allow GASP to intervene in hearings before the board. Though Broughton appeared to carefully separate his professional duties from GASP itself, most likely to avoid accusations of partisanship, Broughton was actually involved in the founding of the group. Broughton had a great deal of contact with GASP, having presented at the same county public hearings in 1969 from which GASP was organized, and subsequently appearing at a GASP-sponsored public meeting on November 17, 1969, and identifying himself as a founding member in early newspaper stories. He certainly moved in the same environmental circles as GASP's members; he was an outdoorsman, climber, and "active in organizing the West Virginia Highlands Conservancy, the North Area Environmental Council, GASP . . . and most recently the Pennsylvania Environmental Council," according to the *North Hills News-Record*. His wife, Suzanne, was also a founder of the North Area Environmental Council of the North Area YWCA, and active within the League of Women Voters' environmental committees.[52]

Bernard Bloom, longtime board member and technical advisor to the group, possessed considerable practical and technical knowledge of air pollution and its control. Born in New York in 1943, Bloom attended college at Washington University in St. Louis, where he encountered the community of activist scientists surrounding Barry Commoner, Ed Lamb, and John Fowler, all of whom he had as professors. Bloom went on to attend graduate school in metallurgy and materials science at Carnegie Institute of Technology in Pittsburgh. "I was in that world of guys—and it was all men," says Bloom now. A dedicated social activist involved in the politics and the social movements of the 1960s, including the antiwar movement, Bloom felt that he didn't entirely fit into the corporate and somewhat militaristic world of materials science research: "My personal orientation was not, get a job, get kids, commute, play golf, buy things. . . . I wasn't countercultural, but I just didn't have that view." After leaving Carnegie Tech, due partly to this mismatch, Bloom completed a master's degree at the University of Pittsburgh's Graduate School of Public Health, studying under Morton Corn and John Frohliger. Bloom began his career as an air pollution control engineer with the Bureau in the early 1970s, and was highly visible both within GASP as a technical advisor, with the Bureau as author of the formula used to create an air pollution index, and in the press as a participant in Variance Board hearings. Bloom quickly went on to positions of national prominence in the EPA as an expert on steel mill emissions, and with multiple local health departments, with aerospace contractors, and as a consultant in indoor air quality.[53]

The organization's professional and academic members chose a strategy based on publicity, the open debate of evidence, expert testimony, and occasional legal action. GASP often portrayed itself to the Pittsburgh public as a brain trust of active professional and academic men, or at least as a facilitator of the interactions of professionals and the public. In support of a November 1971 public hearing, for example, GASP organized speakers to "explain the legal enforcement, medical, economic and technological aspects of air pollution control," including the Allegheny County solicitor, a pathologist, an assistant professor of management science at the University of Pittsburgh, and of course Robert Broughton of Duquesne University, GASP, and the Variance Board.[54] This was a different approach from that taken by the business-oriented Allegheny Conference on Community Development's postwar smoke control campaign, which concentrated on brokering high-level negotiations between industry and government. In contrast, GASP's strategy seemed reminiscent of the academic and scientific debates with which its members would have been familiar: an open discussion based upon expert opinion, the acquisition of evidence, and the presentation of competing arguments. The individuals using this language were nearly exclusively male, and their claims to expertise were pitched as an explicitly rational counterpoint to women's activism.

Just as GASP's primarily female educational wing used rhetoric reflective of the gender of its activists, the professional, scientific, legal, and technical committees of the group employed language emphasizing masculine expertise and masculine concerns. Despite the fact that the group considered itself a mixed-gender organization, the division of labor by gender was clear. As one GASP publication put it, "GASP scientists, attorneys, medical men and volunteers" were responding to pollution problems; there were, of course, no "medical women." At least eleven of the twelve members of the legal committee were male; all of the medical and scientific committee members listed on letterhead in the early 1970s were men.[55]

Messages reflecting responsibility for children were not only directed at women, though such rhetoric still reflected GASP's mission of providing nurturing care for children. For example, a full-page ad in a Pittsburgh weekly newspaper promoted GASP's attempt to get citizen involvement in a pollution hearing, even as it played on a well-known poster from World War II and evoked a paternalistic responsibility for a generation yet to come: "Gee, Daddy, what did you do during the fight for cleaner air?" The father's response: "(Gasp)." The text goes on: "Maybe the fight for cleaner air is a war. Maybe it isn't. In any event, your kids will ask you about it someday. Because what is at stake is their hearts and their lungs. Even more than your own. . . . [T]ake a good look at your kids—ask yourself if it's worth it. Before they ask you." Male GASP volunteers listed their children's health as a reason for activism, as had maternalist women. A GASP publication pointed out that, "As one lawyer put it, he signed on with us so that he can tell his children that he is working for GASP."[56]

Bernie Bloom makes a distinction between the activist and the technical sides of GASP, which may have occasionally broken down along gender divisions, with majority female activists on one side, and majority male technical experts on another. "There were three categories of membership in GASP," Bloom argues, identifying activists who did not have technical knowledge, experts who were occasionally more restrained than others, and then some combination of the two. There was:

> *Michelle, who would not know the details. Then there were a group of people, like Mort Corn and Dan Bienstock, and early on Emerson Venable—then there's shock troops like me, who started as an activist and professionalized myself. . . . Emerson Venable . . . he didn't get along with the activist side of GASP; Mort Corn was the same way, he had private consulting clients, so he had to keep distant from GASP. Venable got himself appointed to the Variance Board. The same thing happened to Daniel Bienstock who got appointed to the Variance Board . . . he had to keep GASP at arm's length.*

Madoff remembers this as a useful division between activists and experts: "Mort [Corn] and I became pals because he sort of used me—and happily I might add—as a way of getting things done. Things that were inappropriate for him to do that I could do." This division between activists and experts was also a gender division, remembers Madoff: "the people who were the mainstay of the group, that made it all happen, were the attorneys and the scientists. And . . . all the universities were basically men and I worked mostly with men when I did any of my research. Very seldom did we have a female."[57]

Reaching the Limits of Local Control

Even while the variance process continued, GASP grew restive with the quality and efficacy of the enforcement of Variance Board decisions, especially considering the still-unresolved matter of the U.S. Steel Clairton Coke Works. In particular, there appeared to be a great deal of confusion over which penalties—county, state, or federal—would be applied when corporations did not voluntarily comply with variance decisions, and what county organization would be responsible for taking offenders to court. GASP member Pat Newman summarized a contentious advisory committee meeting on March 1, 1973, noting that, "The cleanest thing in Pittsburgh right now appears to be the whitewash the County is giving the whole enforcement question, with the Clairton Coke Works mess as the outstanding example."[58]

Briefly put, the authors of the county code had not imagined a situation where a polluter would continue to operate, in violation of county law, without a variance. Here the flexibility of the Variance Board appeared to be too much for GASP to take: "The [Bureau] pointed out that in Article XVIII we only have two kinds of permits: Operating (in compliance) and Variance Operating (under Variance). We collect fees for both," wrote Pat Newman. "However a facility such as [the Duquesne Light] Cheswick [power station] or Springdale Power station belongs in neither category."[59] The binary world of the county code and the shades-of-gray reality of its implementation were not matching up; Cheswick was polluting, but could not be brought into compliance and wasn't about to be shuttered. And that was the problem: how was the county to enforce air pollution codes that, on their face, required noncomplying polluters to shut down? Actually closing down industrial polluters appeared to be beyond the political power of the Variance Board, or anyone in county government.

As the economy declined in the mid-1970s, it became clear that it was impossible for the Variance Board to use its own "nuclear option." It simply could not shut down large, noncomplying industrial sites that employed large numbers of Pittsburghers. Public debate before the board began to include representatives of antienvironmental, pro-jobs citizens groups. In a variance hearing on May 1, 1975, J&L's environmental control manager argued that, if the board failed to approve J&L's request for variance, the "Hazelwood byproducts plant would have to shut down completely or partially." That threat brought support from the president of USW Local 1843 John DeFazio, who said, "We want the petition granted. We don't want to see J&L decline. If you don't grant it, we're going to be in bad shape." This threat was made even more clearly the next day, when a lawyer for a division of Westinghouse Air Brake threatened the immediate loss of twenty-five jobs if the

Variance Board denied its request, and further "raised the specter of the firm moving out of the county." Additionally, Mrs. Jane Kreimer, head of a previously unknown citizens group named Survival with Industry Now through Growth, or SING, appeared before the board to support J&L's request. The similarities with GASP are remarkable, as Kreimer used the public stage created by the board to intervene in the policy debate, even though it is likely, as Madoff observed at the time, that SING had no members beyond Kreimer herself. Still, with labor and a countervailing citizens group taking industry's side, and the threat of job loss made real, the Variance Board appeared unable to force J&L's hand.[60]

As a result of the Variance Board's inability to enforce county ordinance, and as a deteriorating economy diluted public support, the fight over noncomplying polluters moved from county government agencies to county and state courts, but only with extensive pressure from citizens organizations. GASP's role in air pollution control was not yet played out.

• • •

On the whole, it is difficult to evaluate specific implications the variance process had for the process of air pollution control in Allegheny County. While it is true that Articles XVII and XVIII greatly strengthened existing air pollution control regulations and created new enforcement mechanisms, it is also true that the Variance Board was an untested governmental organization. According to GASP, "the whole original point of the Variance Board (to the political and bureaucratic forces who conceived it) was that its existence allowed the County to *not decide*. Therefore we can expect a continuation of the passive role in many cases unless the Bureau is pushed by citizens' groups."[61]

GASP, at least, concluded that the Variance Board had its limits. The county commissioners' choice to distance themselves from political blowback showed an unwillingness to directly engage the entrenched interests of heavy industry, leaving effective enforcement dependent on the vagaries of public support. GASP was in a position somewhere between that of a vigilante crime fighter and a deputized member of a posse: urging the enforcement of the law by a reluctant sheriff, but without the full authority of the law itself. But if the Variance Board was the sheriff, it was still wary of losing the political support of the townspeople with a too-enthusiastic application of the law.

Despite GASP's criticism, the Variance Board was successful in providing a public space to negotiate the complexities of air pollution control, and brought public attention to recalcitrant polluters. While it forged new ground for public involvement in general, it also allowed the direct involvement of activists in the day-to-day process of air pollution control, a previously unheard-of development. The result was an adversarial, antagonistic public process that may have greatly discomfited the representatives of industrial corporations, but at the very least offered citizen environmentalists an opportunity to air their grievances, and certainly tapped into the remarkable collection of technical and scientific expertise available outside of government. More importantly, behind the political cover offered by the Variance Board, county environmental regulators enforced new regulations on the large numbers of industrial polluters who were *not* asking for variance from county code. While GASP and the press focused on the recalcitrance of U.S. Steel concerning coke quenching, countless other technical innovations contributed to a new regime of local pollution control.

Although the Variance Board was discontinued in subsequent revisions of county code, it was replaced with a permitting and appeals process including similar opportunities for public involvement. Thus the give and take of regulator, regulated and public continues today, forty years later. Though GASP might not have counted the board as a great success, the variance process seemed to satisfy the functional and political requirements of county government, distributing responsibility for unpopular decisions and allowing for flexibility in the enforcement of an otherwise stringent air pollution code. The Variance Board's survival throughout this time period attests to its functionality, even if legal maneuverings by Pittsburgh's largest corporations were about to highlight the limits of local control.

6 Citizens and the Courts

UNITED STATES STEEL, JONES & LAUGHLIN, AND THE LIMITS OF LOCAL CONTROL

What did Jones and Laughlin steal? Pittsburgh!

—Woody Guthrie, "Pittsburgh Town"

Federal intervention and a wave of environmental sentiment empowered local environmental activism across the nation in the 1970s, but the actual process of air pollution control was essentially transformed by the substantive inclusion of citizens in the creation and enforcement of new policy. Increased rights for citizen involvement in the air pollution control process made it possible for GASP to intervene in enforcement actions against two of Pittsburgh's largest industrial sources of both pollution and employment: United States Steel and Jones & Laughlin Steel. The narrative of public involvement in the slow and tumultuous enforcement of local law against these corporations demonstrates the transformative importance of new rights for citizen intervention; the continuing analysis of the effect of including the public in implementation and enforcement; the interaction of local, state, and federal power; and yet another instance of the recurring debate over jobs versus the environment. The battle began in 1969 with the substantive inclusion of the public in air pollution regulation, but it escalated when GASP filed civil suits against U.S. Steel in 1972 and J&L in 1975. The struggle continues, unresolved and divisive, to this day.

When Variance Board chair Robert Broughton decided to allow GASP to intervene in hearings in late July 1970, he was not just heralding the beginning of a new, confrontational, adversarial and inclusive phase in local air pollution control, nor simply overturning previous legal precedents of economically determined direct interest. He was also taking action based upon a concept of citizen standing that had previously been much discussed across the nation, but rarely put into practice and never operationalized before a local administrative body. Broughton's decision had wide-ranging implications, and was a turning point in the transformation of air pollution control, empowering the sort of public goading of local regulators that created several decades worth of lawsuits, countersuits, press coverage, and public debate.

The idea that greater public involvement in policymaking could be beneficial was widespread in federal and local politics after World War II. While some federal legislation required and invited public involvement in policymaking by the late 1960s, the public's right to intervene in legal matters was not as clearly defined. Nevertheless, changing currents of legal philosophy on the national level prompted Broughton, the Duquesne University Law professor, to rule against Pennsylvania precedent in 1969 in favor of a largely untried legal theory promoting the rights of the general public to intervene in legal affairs.

In the fall of 1969, Louis L. Jaffe, Harvard law professor and renowned author of a widely used textbook on the relation between administrative agencies and the courts, delivered a paper titled "Standing to Sue in Conservation Suits" at a conference in Warrenton, Virginia. Though the text of Jaffe's paper from the fall of 1969 is not available, it is quite likely that Jaffe described the concept of citizen standing that he had been expounding for most of the previous decade.[1] Put briefly, Jaffe's argument was that all citizens had an inherent, historical right to intervene in legal matters even if there were no direct economic links between individual citizens and the matter before the court, a viewpoint that challenged centuries of legal practice. In the audience that day was Robert Broughton, who took notes on the panel.[2] Broughton's attendance at the conference demonstrates his awareness of the spreading concept and practice of citizen standing, a concept he would soon apply in Pittsburgh.

Jaffe's novel defense of citizen standing and Broughton's application of that concept made possible the substantive involvement of citizens groups in environmental policymaking and enforcement in Pittsburgh. As that legal philosophy spread across the nation, it was operationalized on the local level, institutionalized in federal programs, and supported by Supreme Court decisions in favor of citizen involvement. In Pittsburgh and across the nation, citizen standing made the modern environmental movement possible. In Pittsburgh, it undoubtedly contributed to increasing pressure on the two largest industrial powers, U.S. Steel and J&L.

Pittsburgh's Industrial Landscape

If ever there were success stories to emerge from Pittsburgh's heyday as the industrial center of the world, United States Steel and Jones & Laughlin Steel's would certainly qualify. These two companies were victors in the pell-mell battle of competing and combining enterprises during Pittsburgh's rapid industrial expansion in the late nineteenth and early twentieth centuries.[3] Both companies came to have a massive presence in Allegheny County's economy, politics, and landscape, and employed a large percentage of the county's total workforce while occupying geographically central locations with immense structures. Both were internationally known as powerful examples of American industrial might. Both experienced a long and contentious relationship with local, state, and federal regulators in the years following 1970. By 1975, both were forced into signing negotiated agreements agreeing to limit air pollution. Neither of those agreements, however, meant the end of the story.

United States Steel was created in 1901 when Andrew Carnegie sold his integrated coal, coke, iron ore, and steel manufacturing enterprise to J. P. Morgan. Morgan merged Carnegie Steel with his own extensive iron and steel manufacturing enterprises to form United States Steel, which combined industrial sites throughout the Midwest. The formation of U.S. Steel realized Morgan's dream of bringing every facet of steel manufacturing—from mining raw materials through marketing finished products—under the control of a single, massive corporate structure in order to throttle what he considered to be destructive competition. In fact, as historian Kenneth Warren notes, when U.S. Steel "was formed, it was the largest industrial corporation in the world." On its own, the corporation "produced almost 30 percent of the steel made throughout the world" in 1902. Put another way, U.S. Steel produced between 60 and 65 percent of all rail, structural and plate steel made in the United States in 1901.[4] U.S. Steel's manufacturing sites featured some of the world's largest industrial production centers, including some in the Monongahela River valley that came to feature prominently in histories of labor unrest and environmental disaster. The Homestead, Rankin, Braddock, and Duquesne sites were part of the original Carnegie Steel, but U.S. Steel added the National Tube Works in McKeesport, the Clairton Coke Works (with the largest collection of byproduct coke ovens in the world), a (now-infamous) zinc production plant in Donora, and the Christy Park and Irvin Works. U.S. Steel plants throughout southwestern Pennsylvania and the American Midwest played a central role in earning for the United States the name "Arsenal of Democracy" during World War II by producing untold tons of steel for the war efforts of the allies. U.S. Steel continued to thrive with the postwar demand for durable consumer goods, and in 1969 it was still one of the largest industrial corporations in the United States—and, indeed, the world.[5]

While it is well known that U.S. Steel was a monumental force in the early history of the region, the cultural imprint of the steelmaker continued into the late twentieth century, showing up in the strangest places. To begin with, U.S. Steel had the heritage of the nineteenth century figures of Andrew Carnegie and Henry Clay Frick and the story of the Homestead Steel strike, important to Pittsburgh's labor history. But their cultural impact continued long after the men were gone; Carnegie and Frick's names adorned local libraries, parks, schools, townships, and banks, in a region where vast numbers were employed by U.S. Steel throughout the twentieth century. Stranger still, Pittsburgh's pro football team—the "Steelers"—adopted the "Steelmark" as their helmet logo in the 1960s. The well-known circle-and-hypocycloids image was first created by U.S. Steel in the 1950s and used by the American Iron and Steel Institute in a marketing campaign. The image depicted industry, not sports: each of the different-colored diamond shapes represented a

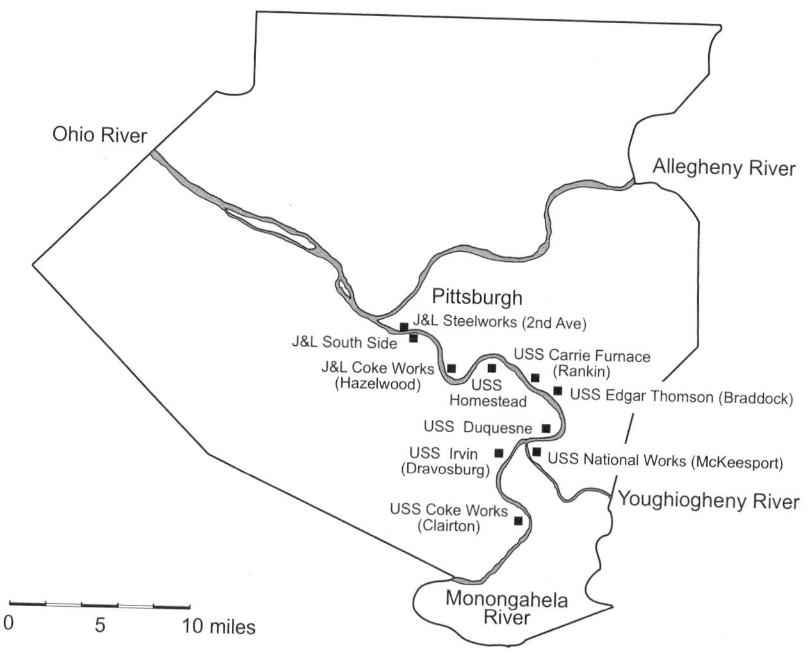

Map 6.1. U.S. Steel and Jones & Laughlin facilities in Allegheny County, 1973.

different ingredient in the recipe for steel, but the team simply changed the word "Steel" to "Steelers" to make the corporate identity of U.S. Steel its own. The most imposing of these cultural impacts was built in downtown Pittsburgh in the 1970s. The U.S. Steel building dominated the skyline with its angular corners and exposed, dark-brown, oxidized steel. At sixty-four stories it was then, and continues to be, the tallest building in Pittsburgh.

Pittsburghers' identification with U.S. Steel was simply a subset, however, of Pittsburgh's identification with steelmaking. This was, after all, a city where the local beer is named "Iron City" and the Yellow Pages are filled with countless businesses named after the "Steel City." Today, the new Steelers' mascot continues this affinity, presenting a civic identity of masculine productivity. In a twenty-first century city where the largest employers are in the fields of medicine and higher education, the football team's mascot wears work boots, overalls, a flannel shirt and hardhat. With bulging forearms, a jutting chin, and the ludicrous name Steely McBeam, the Steelers' mascot demonstrates how Pittsburgh's civic identity is still tied up with ethnicity and masculinity, but also and most clearly with industrial production. The public's relationship with U.S. Steel plays out against the backdrop of this identity.

Among U.S. Steel's many sites throughout the United States and in the Pittsburgh region (as shown in Map 6.1), the Clairton Coke Works was still unique, earning special attention from government regulators and citizen activists due to its size and impact. U.S. Steel built the immense Clairton works during World War I as its first major investment in "by-product" coke ovens, which ingeniously drew off the gases produced by the coking process to heat the ovens themselves and for use elsewhere in industrial processes. The condensed gases produced a rich stew of chemicals and tars, including ammonia compounds, benzene, toluene, naphtha, and xylene. When American access to the products of the German chemical industry was cut off by the war, the "United States had to rapidly create a chemical industry largely based on aromatic compounds" such as those created in the by-product coking process. The by-product ovens thus were more efficient than the previous "beehive" coke ovens, which let most of the gas escape. The new ovens still produced negative externalities, however, since burning the coke oven gas released a great deal of sulfur dioxide into the atmosphere.[6]

In 1970 the Clairton works, located on the Monongahela River several miles southeast of Pittsburgh, was the largest producer of coked coal in the world. Its 1,375 coke ovens produced 22,000 tons of coke from 32,000 tons of coal every day of the week. Each oven was less than two feet wide, but fourteen feet high and forty feet deep, and with 60 or more ovens lined up side-by-side in each of twenty "batteries," the coke works was a formidable physical presence on the landscape. At Clairton, coal for the by-product ovens was washed with water to remove impurities, and then "charged" into sealed ovens from rail cars running along the topside doors. The coal was heated to temperatures above two thousand degrees Fahrenheit for almost a full day to burn off moisture and volatile chemicals, leaving coke, a material that is almost pure carbon. The coke was then "pushed" sideways, out the tall, thin oven doors (amid much dust, gas, and smoke), and transported in specially constructed rail cars to cooling towers. The Pennsylvania Clean Streams Act of 1937 banned the dumping of water (previously used to wash the coal) into streams. Coke producers chose instead to use that "process" water on the cooling coke, producing great billowing clouds of heavily polluted steam and transferring the pollution from one industrial "sink" (the river) to another (the air). By 1970, Clairton was using 2.5 million gallons of process water every day. When the dirty water hit the coke, it released into the air as smoke, steam, and particulates "sixty-seven tons of other pollutants—ammonia, phenols, cyanide, cyanates, ammonium chloride and ammonium sulphate," according to the *New York Times*. It was this process that Article XVII specifically prohibited, requiring companies to clean up process water, to use clean water in the quench, or to control pollutants in steam from the quench.[7]

Like U.S. Steel, Jones & Laughlin Steel Corporation was a physical presence in Pittsburgh, from 1853, when B. F. Jones formed Jones, Lauth & Company. Jones's plan was to operate the American Iron Works inside of the Pittsburgh city limits on the south side of the Monongahela River. When James Laughlin, a partner with

Jones in the American Iron Works, built beehive coke ovens and blast furnaces directly across the river in 1859, it seemed good business to combine the productive efforts of the two enterprises (Jones & Laughlins, Ltd., and Laughlin and Company, Ltd.). Physically connected by a "hot metal" railroad bridge for transport of raw and in-process materials, the two sites on opposite sides of the river combined in 1902 to form the nucleus of the Jones & Laughlin Steel Company, with industrial sites throughout western Pennsylvania. The two Pittsburgh facilities collectively became known as "The Pittsburgh Works," and were eventually joined by the Aliquippa Works in Pennsylvania and the Cleveland Works in Ohio.[8] By the late 1960s, Jones & Laughlin Steel was a billion-dollar-a-year industrial giant, and the Pittsburgh Works was the largest employer in the city. In 1968, J&L was the target of a complex takeover bid by Texas based conglomerateur James J. Ling and his Ling-Temco-Vought Corporation. That bid was slowed by federal antitrust action against LTV, but from 1970 to 1975 LTV was the majority shareholder in J&L, which operated as an independent corporation. In 1984, LTV bought out Republic Steel, merging that company with J&L to form LTV Steel.[9]

Throughout the twentieth century, then, U.S. Steel and J&L were powerful presences in Pittsburgh: fundamental components of the city's longstanding image of itself as an industrial powerhouse, physical presences in the landscape, important parts of the economic history of the area, and two of its largest employers. When it came to the newly ambitious regime of air pollution control, these firms were the eight-hundred-pound gorillas. Their responses to new regulation could spell success or failure for local air pollution control efforts.

The Clairton Saga

Article XVII's attempts to regulate coke production, and its specific prohibition of open air quenching in the coking process, proved particularly difficult for the Variance Board to enforce. With the largest coking facility in the world at Clairton feeding its steelmaking operations throughout the region, U.S. Steel was particularly concerned by the regulation of coking. U.S. Steel applied for variances to continue open air quenching immediately after the new regulations went into effect, and the Variance Board heard testimony concerning those requests throughout late 1970 and early 1971. But no decision was forthcoming from the Variance Board, which found the problem intractable and U.S. Steel less than fully cooperative. In deciding to not make a decision, the board in effect allowed U.S. Steel to continue a prohibited practice without legal action and without a variance. The problem of regulating the complex and dirty coke-making process of the region's largest employers developed into a full-fledged saga, with villains, tragedy, comedy, clashing armies, and no resolution in sight.[10]

While U.S. Steel continued to engage in prohibited practices without an adverse outcome, J&L received a number of one-year variances covering several different sites and practices. But in hearings before the Variance Board, public pressure made it certain that J&L would receive no extensions on its variance from local law. One year was all the time allowed by a Variance Board under the watchful eye of the public. Specifically, J&L made requests for six variances, each covering a number of different sites and processes. U.S. Steel applied for fifteen different variances. Together, these requests dealt with dust produced from mechanical processes to move coal and coke, smoke from coal-fired burners and boilers, and of course coke quenching, all at sites both in Pittsburgh and around Allegheny County.[11]

When the one-year variance on J&L's Blast Furnace Department boilers, By-product Coke Ovens, and South Side Works expired, the company had not taken sufficient action to remedy the original problems, and was thus out of compliance with local law. Still, no legal action was taken against the company, and no extension or variance was issued. Despite that lack, there was no adverse impact upon the company at all. With a lack of decision or enforcement for U.S. Steel's Clairton works and other sites, the Variance Board's attempts to control pollution with J&L and U.S. Steel through 1971 must be accounted a failure. In the case of U.S. Steel, the board had failed to make any decisions regarding major and ongoing violations of existing law. In the case of J&L, they had failed to enforce the law. The members of the Variance Board became restive with the corporate resistance, and in fact began to judge the corporations based not merely upon the actions of the corporations but upon their representative's attitudes and behavior before the board.

By October 1971, Michelle Madoff had blown her top at U.S. Steel, proclaiming her anger in a personal telegram sent directly to the company's director of environmental control (and former chair of the county Bureau of Air Pollution Control) Herbert J. Dunsmore, writing that "we feel it is a little late in the game, after a year of stalling through variances, after a year of fine-free violations, and after the formation of a steering committee to help with the problem, to change directions and lose the faith of the citizens who must breathe your poisons." GASP would soon finish its first film, *Don't Hold Your Breath, Fight for It!* and Madoff promised that the film "will vividly portray the havoc and destruction created by your Clairton operation. . . . [T]he story of your failure to clean up the environment will be told honestly and in detail." She finished her telegram with a phrase describing the corporation's attitude that was often repeated in the coming years: "U.S.S. is again displaying its typical public-be-damned attitude and the citizens will not accept this." This telegram signaled a new public animosity in the increasingly adversarial relationship between GASP and U.S. Steel.[12]

Madoff, though often described as "outspoken," was not providing an opinion very different from that shared by the Variance Board. The seeds of this critical appraisal of U.S. Steel were present early in 1971. After U.S. Steel challenged both the restrictive conditions of a variance for its Edgar Thomson works, and the legality of the Variance Board itself, county solicitor Maurice Louik warned that the board might soon be forced "make a determination of lack of good faith" on the part of U.S. Steel. The adversarial relationship became even more obvious when the board finally ruled on U.S. Steel's petition for a variance covering the use of contaminated quench water at the Clairton Coke Works after nearly two full years of delay. As a front-page story in the *Post-Gazette* noted, in this ruling "the board criticized U.S. Steel for displaying an 'attitude of corporate irresponsibility.'" In the article, unnamed board members characterized the corporation's attitude as "we will not spend money to clean the environment; we will only spend money to bring ourselves into compliance with the law." John Hoerr reported in the *Forum* that public opinion was hardening around the observation that U.S. Steel "has hedged and failed to meet commitments." The *New York Times* picked up the characterization, describing U.S. Steel as "remaining obstinate" in the face of the board's attempts to reign it in. Before U.S. Steel had even come before the courts on its air pollution cases, it had already been branded in the court of public opinion as obstinate, irresponsible, and a recalcitrant polluter.[13]

Once the Variance Board had decided not to award a variance to U.S. Steel, and the Bureau of Air Pollution Control had then determined that U.S. Steel was indeed in violation of local air pollution law, it remained for the county's attorneys to bring suit against the corporation as provided in Article XVII. Unlike all other Variance Board cases, docket number 143 concluded: "10-11-71 Variance Denied. . . . Recommend pursuit in Equity," that is, in court. The appropriate court for the suit was the civil division of Allegheny County's Court of Common Pleas—which combined all local courts into a single state judicial district, and was only one step below the state court of appeals. Representatives of GASP, the EPA, the state Department of Environmental Resources (DER), and the county met informally throughout the winter of 1971 to plan a strategy for the Common Pleas case.[14]

It was fitting that this first major enforcement action focused on the Clairton Coke Works, which received special attention in Allegheny County's air pollution control debate. The works was the only industrial site specifically covered by the lawsuit, despite the fact that U.S. Steel faced ongoing enforcement efforts at a number of other industrial sites. This may have been because of Clairton's status as the poster child for Pittsburgh's industrial polluters. Acknowledging this fact while effectively employing a passive construction, Kenneth Warren writes in his history of U.S. Steel that "[e]missions from the Clairton coke ovens and chemical plants had been a particular source of difficulty." The focus on Clairton most likely stemmed from its physical size; what, really, was the point in enforcing the law on small polluters when the largest coking plant on planet Earth was right next door?[15]

The commonwealth and the county together announced on February 11 that they would file suit against U.S. Steel to force compliance, or failing that, to exact punitive fines. Among a long list of complaints, the suit argued that the county

> *has determined that the defendant has continued and is continuing to operate its quenching facilities in such a manner that it is in continuous violation. . . . This Plaintiff . . . [has] asked the Defendant to refrain from using impure water in its quenching process at the Clairton Works and the Defendant has refused to do so. In addition, Defendant has been denied a variance from the aforesaid regulation of Allegheny County Health Department which deals with quenching of coke.*

In a section claiming that Clairton constituted a public nuisance, the complaint used language reminiscent of GASP's maternalistic rhetoric: the pollution was impairing health "including school children and elderly residents," and that "Primary school children suffer much discomfort from the air pollution arising from the Clairton Works while awaiting school buses."[16] The announcement of the suit was a political affair, with Madoff joining GASP members Pauline Nixon and Arthur Gorr along with Governor Milton Shapp, county attorneys, and all of the Allegheny County commissioners for a joint press conference. The case was assigned to a Common Pleas Court judge with the unlikely name of Silvestri Silvestri, one of ten civil division judges.[17]

Silvestri became an important, though polarizing, figure as the Clairton saga dragged on. One of the few Republicans among the mostly Democratic elected judges and city officials, Silvestri came in for criticism from environmentally minded litigants. His Pittsburgh bona fides were impeccable, however: As a Roman Catholic Italian American who graduated from the University of Pittsburgh for both undergraduate and law degrees, Silvestri was unquestionably a son of Pittsburgh. He was originally appointed to an open position by Republican Governor Raymond Shafer, but was popular enough with the largely Democratic Pittsburgh electorate to manage regular reelection into the 1990s. His decisions in the GASP pollution cases before him put these Pittsburgh allegiances to the test as Silvestri weighed the obvious environmental concerns against the undeniable economic power of old-Pittsburgh employers U.S. Steel and J&L.[18]

While representatives of U.S. Steel, the commonwealth, and the county were marshalling their arguments before Silvestri in 1972, GASP began to actively lobby against U.S. Steel before the public. GASP's film *Don't Hold Your Breath, Fight for It!* was finished just in time to target U.S. Steel, and the focus was clear when GASP took the film on a road show for the members of the House of Representatives in Harrisburg. As the *Press* described it, the "movie . . . 'starred' the belching black smoke from the quenching operations at the Clairton Coke Works of the U.S. Steel Corp. 'Since we cannot take you to the polluters, we brought the polluters to you,' Mrs. [Pat] Wilson [*sic*: Newman] said." As Madoff had promised U.S. Steel previously, "Don't Hold Your Breath" did indeed focus on that company, almost to the exclusion of other sources of pollution. The film shows Variance Board hearings in the Gold Room of the Allegheny County Court House discussing the Clairton Coke Works, along with images of the Clairton Works and of U.S. Steel's Homestead site. After showing rusted signs and children reacting to noxious fumes, the film proclaims that "the enemy is a short distance away on the Monongahela River . . . a coke plant," which, though unnamed, is clearly the Clairton Works. Later in the film, Duquesne University biology professor and GASP member "Manny" Sillman stands on a barren hillside overlooking the Clairton Works and proclaims industry's culpability for the devastated plant life. After the film was shown, GASP president Newman and former president Madoff spoke to the legislature: "It is up to us to save enough clean air for our children—and their children after them," said Newman. Madoff made a politically pragmatic if numerically questionable statement to the assembled legislators: "In Allegheny County alone organized environmentalists are about 350,000 strong and number many hundreds of thousands more throughout the Commonwealth of Pennsylvania. And remember, they vote." GASP member and economics professor Lester Lave answered reporters' questions afterwards, arguing "although Pittsburgh is one of the 10 most polluted cities in the country, 'only $30 per year per person for 5 years' could clean up the air in Allegheny County."[19]

GASP kept the pressure of negative publicity on U.S. Steel with other events. The ongoing debate with U.S. Steel is the context in which a GASP member costumed as group mascot Dirtie Gertie was "unceremoniously asked to leave the Hilton Hotel" after attempting to present the chairman of the board of U.S. Steel with a "special award certificate for contribution to air pollution in Allegheny County" on Valentine's Day, 1972. Edwin Gott was there to receive a "Salesman of the Year" award from the Marketing Executives Association, but security personnel made sure that the costumed Dirtie Gertie didn't make it into the awards ceremony. Even without costumes, GASP and the county were marshalling support for the suit; Art Gorr and Bureau representatives met with the Concerned Citizens of the Mon Valley in April before the suit was officially filed to "see if it has community support." After all of the audience pledged to "do something" and half volunteered to testify in court, one community member presented Gorr "with a jar containing a black liquid which he claimed was from a recent snowfall. 'This is the kind of evidence we want [said Gorr]. . . . It would sure scare the hell out of a judge.'"[20]

GASP kept up the pressure on U.S. Steel throughout June. Drawing upon its members' scientific expertise, Enviro-SOS and GASP collaborated to produce what they called the Task Force Report, a highly detailed document "on the State of the Art, best available technology and conceivable and alternative methods of controlling pollution at the Clairton Coke Works," as Madoff wrote in a letter addressed to U.S. Steel executives.[21] The ten sections of the report covered, among other topics, "Reduction of SO_2 Emissions," "Treatment of Coke Waste Liquor," "Smokeless Charging and Pushing Systems," "Electric Eye Synchronization," and

"Dry Quenching."[22] Each of these sections was the result of collaboration among the members of GASP's Scientific Committee and other local experts, with Madoff and Newman's more critical prose enlivening the introductory section. The technical sections included cost-benefit analyses, discussions of capital improvement financing, collections of pollution measurements, and references to recent articles in trade publications, scientific journals, and academic conference proceedings. An extensive section on "Capital Budgeting" appears be the work of a skilled economist, perhaps Lester Lave, while Edwin Swanson (senior engineer at Westinghouse Electric Corporation's site in Pittsburgh, member of the Pittsburgh-based Air Pollution Control Association, and chair of GASP's scientific committee) provided technical information on the feasibility of having Clairton wastewater treated by Allegheny County's Sanitary Services, a solution that held promise early on in the Clairton saga but ultimately foundered. Perhaps the most eye-opening component of the report was GASP's use of an internal U.S. Steel memorandum from 1968 to establish amounts of contaminants released daily by process-water quenching: forty tons of ammonium chloride, four tons of phenol or carbolic acid, six tons of cyanide and thiocyanate, and seven tons of oil and other impurities. U.S. Steel subsequently denied the existence of this internal memo, which GASP dubbed "the purloined letter" and proceeded to reproduce in many of its publications.[23]

Madoff hand-delivered a copy to U.S. Steel, where some were beginning to feel a bit beset:

> *Well, when we wrote the report about the Clairton Coke Works, I called [U.S. Steel vice president]Earl Mallick to give him a copy. . . . And I went up to the building to deliver it to him and went to the 16th floor [of the U.S. Steel tower] . . . and a bunch of men were sitting there, and they looked at me and I looked at them, and as I walked down the hall, I heard one say, oh, my God, we're not even safe in our own building.*[24]

Even if U.S. Steel wasn't appreciative, others were thankful for GASP's efforts in creating the Task Force Report; Judge Silvestri Silvestri called it "an extremely valuable source" that "was of assistance to me in understanding the nature of the problem with which we had to deal." The law firm hired to prepare the county's case against U.S. Steel called it an "invaluable tool." The director of the state DER's Environmental Pollution Strike Force wrote, "Without your Task Force Report, the preparation of a settlement proposal adequate to protect all of the people of this County would have been much more difficult and perhaps less authoritative. . . . If every similar regulatory agency had a group on which it could depend for technical information such as GASP, the fight against pollution would be much easier." The Task Force Report was lauded in the local press, not least for its ability to speak the language of technical expertise. As the *Forum* put it, "GASP delivered a rather remarkable uppercut to the stuffy paunch of USS—a 30-page, 10-section report as technically detailed as though it were prepared by Steel's best experts." This was clearly substantive public involvement in policy implementation and enforcement; the citizen environmentalists had pushed their way into the policy process and made a meaningful contribution to shape the results to their ends.[25]

Even while U.S. Steel was actively involved in fending off pressure from labor unions, politicians, and regulators over sites from Gary, Indiana, to Homestead, Pennsylvania, Judge Silvestri was directing the parties before him in the Court of Common Pleas to negotiate a consent decree over Clairton.[26] The decree would be a consensual agreement between the parties that, when signed, would have the force of a court order and could be administered and managed by Silvestri. In late August, when news of successful negotiations leaked to the press, it was seen as a ringing victory for air pollution control, with special accolades reserved for the beneficial involvement of citizens. In lengthy *New York Times* coverage, E. W. Kenworthy wrote that "The State of Pennsylvania, Allegheny County and an aggressive environmental organization named GASP have demonstrated that local government with citizen support can force an industrial giant as big as the United States Steel Corporation to adopt an antipollution program far beyond anything it had planned." The negotiations produced three separate consent decrees, on the topics of coke ovens, sulfur reduction, and process water for quenching, all signed by the end of October.[27]

The coke oven consent decree was a highly detailed and technical ten-page legal document signed by county commissioners, representatives from the state attorney general, and U.S. Steel's lawyers from the firm of Reed, Smith & McClay. Among many other requirements, section 1(d) of the decree required that coke oven doors be sealed so that they could not leak any smoke for any period after the first fifteen minutes, while section 1(e) required that "[o]n or after December 31, 1972, there shall be no visible emission, except non-smoking flame, from any opening on the coke oven doors from more than ten percent (10%) of the coke ovens in any battery at any time." Section 3 required U.S. Steel to "submit to plaintiffs an air pollution abatement plan" by September 1973 describing specific steps to meet the many emissions limitations of the decree. In exchange, U.S. Steel would limit its legal liability, since section 9 specified that "Plaintiffs shall not initiate any legal proceeding or prosecute Defendant under any statute, rule or regulation of the Commonwealth or the County should Defendant violate any provision of the decree," leaving all legal oversight up to Silvestri. Section 13 further decreed that the U.S. EPA sign off on the consent decree as an approved part of the Clean Air Act State Implementation Plan.[28]

But by far the most far-reaching of the consent decree's sections was a complicated immunity deal for U.S. Steel: "In the event the Defendant fully complies with all provisions of the approved plan but, nevertheless, is unable to meet the final standards . . . it shall be relieved of full compliance with said standards for a period of ten (10) years." The requirement for EPA approval, combined with the ten-year amnesty clause, seemed to guarantee U.S. Steel a waiver from county, state, and federal regulations if emissions standards proved impossible to meet.[29]

While local political officials supported the pact—all three county commissioners, Leonard Staisey, Thomas Foerster, and William Hunt, had personally signed the decree—the document depended upon EPA approval to become enforceable. The local press and politicians supported the voluntary nature of the agreement, and assumed EPA support of a locally negotiated settlement would be automatic. After all, as the editors of the *Post-Gazette* argued, "under a consent decree all parties have a stake in making it work. This is preferable to a court order where progress might come only grudgingly."[30]

There seemed to be a sense that the consent decree was the beginning of a functional solution for Pittsburgh, and possibly for the nation. Despite the concessions for U.S. Steel, "even the most militant . . . considered these to be acceptable compromises," wrote John Hoerr in the *Forum*. As he saw it, Clairton was too big a train to turn quickly: "The reality of the situation, however, is that even if USS were to start tearing down its coke batteries tomorrow, it could not possibly replace its current cokemaking capacity" anytime soon. And, wrote Hoerr, "No matter how mad people get about U.S. Steel's failure's of the past, no sane environmentalist wants to vent anger against 30,000 steelworkers and their families." GASP and others supported the decree as an acceptable compromise in an extreme situation, and this collaborative spirit of negotiation seemed worthy of praise on the national level. The *New York Times* wrote that the decree represented "a considerable victory for the plaintiffs and for clean air."[31]

Despite the support of local interests, it quickly became obvious that the EPA would oppose the parts of the decree that gave U.S. Steel immunity from prosecution. After an August public hearing on the matter, EPA regional administrator Edward Furia visited Pittsburgh for a poorly received whirlwind tour of the Clairton site. As the local press later described it, "At times, the EPA visit resembled a three-ring circus with more than 30 newsmen, TV cameramen and radio personnel milling around in wild confusion." After speaking to the press, Furia made a surprise visit to the coke works, overflying the site in a helicopter along with a few television news cameramen, and landing only to denounce the proposed consent decree as "blackmail." Furia flatly declared that the portions of the decree that guaranteed immunity from federal prosecution were "unacceptable."[32]

All local stakeholders in the consent decree closed ranks against the unwanted federal intrusion. The three county commissioners who had signed the decree were infuriated by Furia's tour. William Hunt declared: "In the strongest terms I have suggested that the Philadelphia administrator [Furia] be withdrawn from this area and replaced with someone who has the expertise to make a proper decision, removed from all thoughts of self glorification."[33]

GASP, whose interest might have been thought to lie with the EPA, in fact joined local politicians in opposing EPA intervention, and was an outspoken critic of Furia's actions jeopardizing the agreement. Madoff remembers his actions now as "a grand stand. We never did that . . . Furia flew into Pittsburgh, landed on the top of the U.S. Steel [Tower]. . . . And he was going to take some kind of definite stand. Well, he should not want to close down U.S. Steel. That wasn't the game. It was not to close down jobs. It was to get them to comply. And he needed to come in with something that wasn't 'we're gonna shut you down' on the top of their building." GASP defended the consent decree so strongly that it considered it an honor when Commissioner Foerster presented the group with the pen used to sign the consent decree for GASP's third birthday. The inscription thanked GASP and Madoff for their support "in arriving at an effective and equitable pollution control agreement." While political scientist Charles O. Jones believes that GASP and local politicians banded together to defend their consent decree as an aspect of a somewhat nativist rejection of federal intervention, it is also possible that GASP was defending the process of local air pollution control, not U.S. Steel's particular case. After all, GASP had invested a great deal of time and effort in creating and supporting the county's mechanisms for air pollution control, and federal intervention might mean those mechanisms would falter. GASP members' interest as citizen environmentalists was to maintain the process of substantive involvement in all instances, not just the Clairton case.[34]

Despite the brouhaha over federal intervention, local and national press hailed the completed consent decrees as hallmarks of success and progress in the battle to control air pollution. Accompanied by a dramatic photo of the Clairton works pumping out thick black smoke, *Newsweek* wrote that "For more than 30 years, U.S. Steel's Clairton coke works has been fouling the western Pennsylvania air with polluted steam from the quenching of its red hot coke," but the consent decree, negotiated "in response to a suit filed by the Commonwealth of Pennsylvania and Allegheny County, and at the urging of a citizens' organization called [GASP]," promised an end to the pollution. In response to journalistic celebration of the plan and evidence of broad public support from Pittsburgh's two hearings on the issue, the EPA worked out a face-saving compromise whereby the version of the consent decree presented for EPA approval did not include the immunity clause. In any case, EPA administrator William Ruckelshaus claimed that the clause was beyond his legal authority. A *Post-Gazette* editorial titled "Calming the Clairton Fuss" credited Ruckelshaus for going "about as far as an administrator could go in all but apologizing publicly . . . for impetuous actions of his subordinates."[35]

By the first days of 1973, the decrees were accounted a success, as GASP noted in its own publications:

> *The consent decrees mark a turning point in the battle against pollution and the beginning of a major cleanup in the Monongahela Valley. In the next year and a half residents of Clairton, Liberty, Portvue, Lincoln and Glassport will see real improvement in air quality, with smoke from Clairton coke ovens sharply reduced. In 1975 we expect to have clean air in the Clairton area for the first time in half a century.*

This was clearly the success story that all sides wished to tell: a voluntary compliance schedule worked out (relatively) amicably between all sides in the local arena, without resorting to federal intervention or to long, drawn-out court battles. The hope is almost palpable that citizen environmentalists and corporations had found a pragmatic and practicable common ground through rational negotiation and compromise, providing a successful model not only for resolving Pittsburgh's problems but also possibly the nation's. As 1972 came to a close, none of the local actors could have known that the consent decree actually marked the beginning, not the end, of the conflict. It is only with hindsight that the irony becomes clear: the consent decree was wildly popular immediately before it became wildly unpopular.[36]

These agreements, so laboriously worked out and defended by local interests, fell apart in a matter of months when it became clear that U.S. Steel could not, would not, or never meant to fulfill its side of the agreement. The first complaints came from a citizens group loosely affiliated with GASP: Concerned Citizens of the Mon[ongahela] Valley. GASP member Pauline Nixon was also a member of the Concerned Citizens, and the members of that group began to publicly register their complaints about continuing Clairton pollution soon after the beginning of the new year. As GASP later argued, "terrific public indignation" arose to pressure the county to hold U.S. Steel to the requirements of the consent decree.[37]

The coke oven consent decree was not the only component of Pittsburgh's air pollution control debate concerning U.S. Steel in the spring of 1973, but Clairton dominated other stories. When a series of events seemed to indicate that U.S. Steel was reneging on the laboriously negotiated deal, the reaction was swift and harsh when compared to other matters involving U.S. Steel. While public complaints from Concerned Citizens of the Mon Valley began in early January, momentum was added to the issue when Pittsburgh suffered another of its periodic air pollution emergencies on January 18. Finally, a critical report on U.S. Steel's noncompliance from Bureau engineer (and GASP member) Bernie Bloom provided expert evidence to support the public's hunch: U.S. Steel was not living up to its end of the bargain. In an ACHD internal document which ended up in GASP's hands, Bloom, head of the Bureau's "Coke Plant Inspection Group," wrote to the county's legal counsel that "U.S. Steel has violated [Section 1(e) of the coke works consent decree] on the vast majority of our observations. . . . Clearly most batteries are in *continuous* violation." The report concluded that sections 1(d) and 1(e) of the decree "have been violated respectively, during 95% and 91% of our observations. U.S. Steel is aware of our observations. I recommend we return to court for enforcement."[38]

On March 27, the state and the county, as signatories of the consent decree, petitioned Judge Silvestri for an enforcement decision against U.S. Steel. The petition specifically argued that U.S. Steel was violating sections 1(d), 1(e), 2(e), and 10. These provisions of the agreement directed the company to seal coke oven doors to stop emissions, prohibit visible emissions from more than 10 percent of the coke ovens in a battery at any time, prohibit emissions from the top of the ovens, and in general to "maintain and operate the Clairton Works at all times hereafter in such a manner that the air contaminants therefrom are minimized to the greatest extent possible."[39]

In response to this petition, Silvestri held six days of hearings in mid-April, during which the company argued that despite its best efforts, it was unable to meet the requirements of the consent decree it had signed because it believed that the requirements of the decree were fundamentally impossible to meet. In response, county attorneys argued for a contempt of court charge coupled with a $300,000 fine. The hearings eventually moved to the Clairton works themselves for their final day, when the "Judge donned a hardhat, a respirator, goggles and protective clothing for a random survey of the world's largest coke plant." The state DER's Bureau of Litigation Enforcement director Marvin Fein had by this time ratcheted up the rhetoric, declaring that if emissions continued, "I recommend we put the company officers in jail."[40]

Judge Silvestri's decision on the petition, issued on May 23, displeased all sides: the court ruled that although "there was more than sufficient evidence to support [the] conclusion that USS was in violation of all four" sections of the decree, that U.S. Steel should *not* be held in contempt for violation of the consent decree, and "specifically stated it was not imposing any sanctions for violations." As in the original creation of the consent decree, Silvestri demonstrated a desire to negotiate a voluntary settlement, choosing to set up a joint evaluation of the original consent decree to determine if, as U.S. Steel claimed, specific requirements of the decree were impossible to achieve.[41]

Dissatisfaction with this reexamination of the terms of the consent decree—which local politicians had negotiated and then vehemently defended only six months earlier—resulted in a variety of actions from many organizations. State representatives formally petitioned the decision to Silvestri, who initially refused to reconsider his decision. At the same time, the United Steelworkers announced "it was 'considering' a request to EPA to take legal action." Using increasingly legalistic language, GASP escalated its pressure: GASP attorney Arthur R. Gorr petitioned the county commissioners to intervene in the matter on June 7, arguing that Silvestri's actions were "tantamount to a judicial nullification of Article XVIII, a code duly enacted by the county, and that the county is duty bound to seek a reversal of this action by appeal to a higher court." Following GASP's urging, the county commissioners voted on June 10 to appeal to Silvestri for a new hearing. Finally, the state and county together petitioned Silvestri for an opportunity to testify before him on the matter.[42]

The matter was quickly spiraling outside of the comprehension of even a reasonably informed member of the community. As the *Press* put it, "Common Pleas Court Judge Silvestri Silvestri has reconsidered his decision not to reconsider his original decision refusing to penalize U.S. Steel for failure to limit air pollution at its Clairton coke works. If that sounds confusing, it is." In the end, despite the pressure exerted by county and state officials, Silvestri announced on June 25 that his original ruling would stand: U.S. Steel was out of compliance with county law, but would not be punished. This was a return to the status of the matter more than a year previously, when the Variance Board appeared unable to force compliance.[43]

In response, GASP took the unprecedented step of preparing its own civil suit against U.S. Steel on June 25. In a letter sent to the EPA, U.S. Steel, the governor of Pennsylvania, and nearly every local politician or bureaucrat involved, GASP's legal committee chair Arthur Gorr wrote that GASP intended "to file a civil action against the United States Steel Corporation under section 304 of the Clean Air Act." In other words, GASP was preparing a citizens' suit enabled by federal law, the first of its kind in Pittsburgh.[44]

Art Gorr, GASP's official legal counsel and eventual president, was the driving force behind GASP's civil suit. Still a junior partner at the small Pittsburgh law firm of Stein & Winter, Gorr practiced criminal law. But he appears to have leapt at the opportunity to work on Pittsburgh environmental issues, drafting countless letters on legal matters for GASP on Stein & Winter letterhead. Gorr was particularly angered by U.S. Steel's violation of the consent decree they themselves had originally negotiated and signed, or any suggestion that the negotiated consent decree was unenforceable. In a 1973 letter to Ronald Chleboski, head of the Bureau, Gorr wrote that he was "furious with the suggestion that certain county officials have doubts that U.S. Steel can meet the standards and are now considering negotiating the decree." Gorr also wrote to Bernie Bloom and county attorney Gerald Dodson that a *Press* article that implied the consent decree could be renegotiated "annoyed the hell out of me." He continued: "U.S. Steel agreed to a certain standard; it has not met that standard; the burden is on U.S. Steel to explain why not; the burden is on U.S. Steel to come forward" and show what it had done to comply.[45]

While Gorr was motivated to volunteer his time to work on GASP's lawsuit pro bono, he faced a considerable challenge: This would be GASP's first lawsuit, the first citizens' suit under the Clean Air Act in the Pittsburgh area, and the first prosecution against industrial giant U.S. Steel for pollution violations under new county laws. If that were not daunting enough, Gorr had to know that he was challenging the formidable resources of Reed, Smith & McClay, the large, Pittsburgh-based law firm that had represented both U.S. Steel and Jones & Laughlin in legal matters for nearly a century.[46]

But GASP was committed to the legal action. Even before it had filed its own suit, GASP's leadership felt that legal action against U.S. Steel was symbolically important, and could "discourage others from using legal time to avoid compliance." GASP's denunciations of U.S. Steel, along with continued local political support for enforcement, led to a joint state/county appeal of Silvestri's decision to the Commonwealth Court in November 1973. While this was a positive step in GASP's view, the appeals process was a lengthy one. The Commonwealth Court would not even hear the case until May 1974. In the meantime, U.S. Steel continued to operate its coke works in violation of county law, without a variance, and in breach of its agreement in the consent decree. Despite this, U.S. Steel was not subject to punitive fines, closure, or legal punishment of any sort. Still, litigation seemed the only viable option for GASP: "We may lose another year," admitted Michelle Madoff, "but we'll gain in the end with a final, irrevocable decision. What we've got now is borscht. We've got dishwater."[47]

The U.S. Steel issue regained its central importance for the local air pollution debate in September and November 1973, when the corporation's plan to reduce air pollution was rejected by the state, prompting federal authorities to intervene in local matters. One of the requirements of the 1972 Consent Decree was that U.S. Steel would develop a long-term air pollution abatement plan, which was delivered to the state on September 25. Both the state and the county rejected that plan as not satisfying the conditions of the decree.[48] This appeared to be the breaking point for GASP. Though it had previously opposed the intervention of the EPA in the form of Edward Furia, GASP president Kitzes now wrote to EPA Region III director Daniel J. Snyder to ask for federal legal action against U.S. Steel. Indeed, the EPA gave a thirty-day notice of intent to sue U.S. Steel. This did not actually lead to the filing of a suit thirty days later, as GASP noted: "EPA Region III gets into the saga when on November 9, 1973—with the issuance of what must have been the longest 30-day notice in history—it ordered USS to clean up its particulate pollution at CCW." The intent letter led instead to "formal negotiations" between U.S. Steel and the EPA, with U.S. Steel vice president Earl Mallick publicly questioning the legality of federal action while state negotiations continued.[49]

Continued negotiation was insufficient for both GASP and the state DER; GASP, along with Michelle Madoff's new organization CETF (later Enviro-SOS) and other groups, "called on the state, county and federal governments to join forces against U.S. Steel Corp." Madoff, as always, went further, calling for possible "jail sentences for those who commit crimes in the board rooms of U.S. Steel." In particular, the citizens groups urged all regulators to dissolve the original consent decree, which had required U.S. Steel to submit an acceptable cleanup plan, on the grounds that the corporation had produced an unacceptable plan. As *Press* conservation editor Fred Jones noted on November 28, "In a bitterly worded letter" the state DER announced that it was rejecting the U.S. Steel cleanup plan, as "the company has shown a complete lack of good faith in the cleanup effort of Clairton and accused the steel firm of a callous disregard for the health and welfare of Clairton area residents."[50]

There was slow progress on the matter early in 1974, as the state was convulsed in a political battle that threatened to render the Clairton matter moot. For several months, GASP and others across the state waged a public relations battle against a set of proposed laws, known as the "Bethlehem Bills," which "would give industry the right to discharge pollutants into the air and water of the Commonwealth without regulation or penalties for a period of 12 years," according to the *Press*. The idea was that declining industrial facilities—like Johnstown's Bethlehem Steel, represented by state legislator John Murtha—would be exempt from all pollution regulation as long as they stated that they were closing sometime during the next twelve years. The bills specifically stated that "citizen groups would be barred from seeking relief in the courts," limiting appeals to "any person, partnership, corporation or association having a financial or economic interest which is or may be affected." In effect, the Bethlehem Bills reinstated direct interest as a requirement for standing, and according to the state's Marvin Fein, "would set back Pennsylvania's air and water pollution controls to the 1920s." Negotiations between the state DER and Bethlehem Steel eventually made passage of the increasingly unpopular bills unnecessary. Still, the Bethlehem proposal demonstrated increasing economic concerns about environmental controls.[51]

Other matters were in progress, of course: U.S. Steel was negotiating before the Variance Board for its Edgar Thomson, Duquesne, and McKeesport plants, even as the county and GASP called for fines and Michelle Madoff "accused the firm of dragging its heels." This resulted in a cleanup plan, announced in June. Despite that limited success, the language used by county regulators was becoming increasingly critical of U.S. Steel in matters other than Clairton. Chairman of the Variance Board Anthony Picadio indicated that he feared the Variance Board was "getting had" by U.S. Steel, while county solicitor Gerald Dodson "questioned U.S. Steel's sincerity" and Michelle Madoff flatly declared that the company was only pretending to follow the law as "another example of the 'fine art of buying time.'" Discussion of the Bethlehem Bills may have contributed to increasingly negative depictions of industry, with the *Post-Gazette* decrying "selfish industry" and the *Press* summarizing the opposition to the bills as an attempt to "emasculate" the state's regulatory control and "give industry the right to pollute as they pleased."[52]

But the slow grind of the courts was continuous on the matter of Clairton, if largely submerged from public view. The first development broke the surface on July 12, when the Bureau brought suit for an additional $102,000 in fines against U.S. Steel.[53] But that was quickly overshadowed by even bigger news regarding the stalled appeal of the 1972 consent decrees. In September 1974—nearly five years after U.S. Steel first appeared before the Variance Board, and two years after the original consent decrees were signed—the Commonwealth Court published its decision. The seven judges of the court unanimously ruled that U.S. Steel could *not* argue that the original consent decree was impossible to satisfy, and furthermore that Silvestri could not change the decree, because U.S. Steel voluntarily entered into what was, in effect, a binding contract. As the court opinion read, "at this point in these proceedings, it makes no difference how distasteful the consent decree may be to USS. The fact remains that USS voluntarily entered into this agreement with the Appellants, and the agreement was approved by the lower court." The judges took issue with the legality of Silvestri's attempts to reopen the decree, though they appeared to appreciate his intent: "The [lower] court was in error in attempting, *sua sponte* [on its own], to conciliate the matter or to determine whether the consent decree was in need of modification. This is so in spite of the laudability and common sense of such an approach."[54]

GASP immediately pressed the Bureau to begin contempt of court prosecution against U.S. Steel for violating interim standards in the original consent decree, which it did on December 4, though Silvestri would not take up the matter until early the next year. Together, the state and county now were asking for $3 million in judgment against U.S. Steel. In the meantime, possibly because the issue was not complex enough, a federal grand jury was empanelled to begin prosecution of U.S. Steel's violations of the Clairton Consent Decree under the Clean Air Act. Beginning in October 1974, the grand jury subpoenaed the officers of U.S. Steel, which appealed on the grounds that federal prosecution should not be allowed to continue while state and county action was already ongoing. The grand jury was delayed on appeal for a year, thus beginning a long history of federal court intervention on this issue, which thankfully is not the subject of this work.[55]

Back in Pittsburgh, mounting legal pressures on U.S. Steel spurred on public denunciation, and GASP poured salt on the wound. On November 20, 1974, it gave the Dirtie Gertie award to U.S. Steel for the fourth year running, "to 'salute' our major air polluting industries in Allegheny County whose irresponsibility towards the health of the community and lack of cooperation with government clean-up regulations and programs" had made the Pittsburgh area such a highly polluted environment.[56]

Despite the Commonwealth Court decision, on January 20, 1975, U.S. Steel petitioned Silvestri to declare the previous consent decree null and void, and to eliminate all fines associated with violations of the decree, as the company could not meet the requirements of the decree. The state and county disagreed, arguing that the consent decree could not be set aside without their approval, and that U.S. Steel had not made a good-faith effort to meet the requirements and thus should not be allowed to ignore them. Judge Silvestri heard extensive argumentation on the matter throughout the spring of 1975, including depositions, expert testimony, and photographic evidence. In an action that deeply embittered the citizen environmentalists and representatives of the commonwealth and county, Silvestri announced in June that the consent decree would indeed be set aside, despite the Commonwealth Court's 1974 admonition on the matter. His decision on the matter defended his jurisdiction under the original consent decree: "This case is predicated on a consent decree a fact which the plaintiffs acknowledge by their continued reference to the same in their lengthy and detailed brief. However, the plaintiffs studiously avoid paragraph 11 . . . which provides: 'This decree and any plan or schedule incorporated herein may be modified by agreement of the parties or by this Court upon petition by any party.'" In short, Silvestri wished to exercise his power to reopen negotiations on the original consent decree and force a new agreement.[57]

The hope of 1972 was entirely dashed by 1975. By July of that year, the *New York Times* was calling U.S. Steel the "*bete noire* of E.P.A.'s enforcement office," and arguing that the long, drawn-out legal battle, "a Dickensian horror" in the case of Clairton, was an official U.S. Steel strategy to avoid compliance. After praising the original consent decree several years earlier as a desirable alternative to arbitrary regulation or punitive litigation, journalists and politicians had nothing but contempt for a decision that set aside that earlier success as fatally flawed.[58]

It became evident that the commonwealth was particularly frustrated by developments, as state attorneys filed a petition to have Silvestri removed from the case, accusing him of bias and of accepting campaign funds and support from attorneys at Reed, Smith & McClay, which represented U.S. Steel. Silvestri referred the petition for his removal to President Judge Henry Ellenbogen of the Common Pleas Court even while the court elections were under way that fall. Ellenbogen inherited both the political mess of one of his judges accused of bias and the practical mess of the failed consent decree. The recusal petition itself was a political hot potato: while the state's attorneys seemed intent on removing Silvestri, the county commissioners were not, angrily repudiating assistant county solicitor Gerald Dodson's signature on the petition. *Press* columnist Roy McHugh reported that "the Republican County Commissioner, Dr. William Hunt, recommends chopping off [Dodson's] head."[59] Silvestri spoke to the press in his own defense, declaring the charges a "smear" campaign by the state. With apparent frustration, Ellenbogen issued a gag order on all of the involved parties, including Silvestri, so as not to affect the ongoing elections. His subsequent decision attacked the state's lawyers, making much of the fact that the county did not join the complaint, and implying that by serving Silvestri with papers "at his home late in the evening of Sunday, September 14," the state was ambushing an honorable man. As he addressed each of the objections in turn, Ellenbogen hotly defended his colleague, and questioned the state's intentions. In the matter of campaign contributions, Ellenbogen wrote that "We find it incredible that the Commonwealth seriously asks us to believe that a judge of this court would sell out for such a paltry sum." Ellenbogen's decision cleared Silvestri of all allegations, and as a parting slap, placed court costs on the state.[60]

Even with Silvestri nominally returned to control, Ellenbogen took charge of the consent decree negotiations himself, marking by his actions the fact that Silvestri had lost the ability to serve as arbiter. Madoff remembers Silvestri with anger: "Silvestri was a two-faced son-of-a-bitch," she says now. Many of the citizen environmentalists involved retrospectively question Silvestri's attempts to mediate a consensus position, despite Ellenbogen's decision. As Madoff puts it, "[Henry] Ellenbogen took over and Silvestri was taken off the case. Why would that have happened? One must question that."[61] The growing tension led GASP to distrust Silvestri. Art Gorr was careful to keep his distance from the Clairton hearings after Silvestri momentarily jailed Gorr for contempt in an unrelated personal injury case earlier in the year. Gorr appealed the charge to the state Supreme Court, with the support of the county bar. Between Silvestri's actions, the state's charges, the county commissioners' criticism, and Ellenbogen's backlash, the Clairton case was quickly devolving from technocratic regulation to courtroom soap opera.[62]

Despite the kerfuffle, by 1976 GASP, Allegheny County, and U.S. Steel seemed to be back where they had started six years previously, when the Variance Board was unable to force compliance. U.S. Steel was operating in violation of local law, but no enforcement of that law seemed possible. The process of negotiating the enforcement of local air pollution laws was so far from being done that it had begun again, as if from square one. In Ellenbogen's words, "Long, exhaustive, and detailed negotiations" in his chambers resulted in 1976 in a completely new consent decree, replacing the 1972 agreement and promising that Clairton would be clean by the far-off date of 1983. U.S. Steel would not be operating in compliance with county laws, but neither would it be paying the more than $3 million dollars in fines it had accrued in its 241 violations of county code, though it voluntarily donated $750,000 to local educational institutions to study environmental problems in the county. Crucially, the "exhaustive" negotiations organized by Ellenbogen were closed to citizen environmentalists and to the press: "Secret Meeting on Clairton Coke Works Held Here" was the headline in the *Post-Gazette* on September 25, 1975; only one lawyer was allowed into the "two-hour closed summit of the principals in the $3 million contempt action." The principals, who were barred from speaking to the press by Ellenbogen's order, included the county commissioners and representatives from U.S. Steel and the USWA, EPA, and state. Neither GASP nor any other representative of the public was allowed to intervene. It seemed as if the process of air pollution control had reverted back to its corporatist phase, excluding the public from closed-door, upper-level negotiations: "Fearful of Being Ignored," read the *Post-Gazette* headline in 1976, "GASP Wants Voice on Clairton Air Pact." It might as well have been 1966.[63]

The Language of Job Loss, U.S. Steel, and Jones & Laughlin

GASP's experience with U.S. Steel shaped its relationship with the region's second largest industrial corporation, Jones & Laughlin Steel. As a result of its increasingly adversarial relationship with U.S. Steel, the citizens organization was much more ready and willing to take J&L to court when that corporation appeared to be following U.S. Steel's example in delaying compliance with local law. The skills of legal intervention, acquired over several years in the battle with U.S. Steel, were immediately applied in the case of J&L. But the continuous threat of job loss, present in all environmental debates with large employers in the Pittsburgh region, seemed to trump other concerns in the J&L case, perhaps because matters were only elevated to the courts in the mid-1970s, as economic conditions declined.

Unlike U.S. Steel, J&L did receive permission to operate all of its industrial sites under variance from Article XVII in 1972. But when those one-year variances ran out in 1973, the company had not substantially altered the prohibited practices, leaving it in violation both of the original Article XVII and the new XVIII. Immediately, GASP pressed for enforcement action, arguing that small polluters were obeying county law while large polluters were defying it. As Pat Newman put it, "In our County's Rogue's Gallery of Vicious Polluters there are the large, the small and the very small. As pointed out before, it is generally the very small which are hauled into court and fined and made to quit. The others continue."[64]

After a year of negotiations the Bureau filed suit against J&L for $100,000 in fines on July 2, 1974. GASP was quick to file a motion with the court to appear as an intervenor in this case. But the discussion of J&L's compliance with local law was quickly derailed into extended discussion of possible job losses in a worsening economy. Madoff had a hand in this, preemptively predicting that if J&L chose to shut down some operations at its Pittsburgh works, "it will blame environmentalists for the job losses involved," and that "Environmentalists are sick and tired of being made the scapegoats when a company decides to shut down operations for its own selfish reasons." A month later, as discussion of the possible sale of J&L to LTV picked up, Madoff reiterated this theory, predicting that the works would be entirely closed before more strict air pollution laws came into effect in 1978, "posing the loss of 6,000 jobs." But this language really took off when used by those who were truly in fear of losing their jobs. In the next year, even as fears of the Pittsburgh Works' closure by LTV faded, local unions began using job loss as an argument against any environmental regulation. "Take J&L away and you can have all the clean air you want, but there'll be no more jobs," one USWA local president warned the Variance Board.[65]

On March 24, 1975, the front page of the *Press* proclaimed "GASP Sues J&L on Pollution Control," noting that "GASP said this is the first time a citizen's group has taken action against a polluter for existing violations." Indeed, GASP's press release called the action "unprecedented," and noted that the suit called for enforcement, but asked for the court "to award GASP only costs to cover the pending litigation." Along with the EPA's announcement of 115 infractions of the Clean Air Act at J&L plants, the GASP suit forced J&L back to negotiations, resulting in a consent decree that would have J&L spend $200 million on environmental controls, while keeping J&L's "6,000 jobs" in the area.[66]

The strongest argument for corporate interests clearly lay in the perennial debate of jobs versus the environment, and concerns over plant closings or layoffs entered the public discussion more forcefully in 1975 than at any other time since the creation of Article XVII. For steelmakers, who could in theory leave the Allegheny County area if local regulation of their activities grew too onerous, plant closure was the ultimate bargaining chip. This of course had been present in almost all discussion of such matters. Robert Broughton had nodded to economic pressures in a Variance Board decision in October 1970: "While employment is not a justification for polluting the air, we should take cognizance of the inequities involved when the cost of solving a problem, for the benefit of the community at large, falls disproportionately heavily upon a few people." But this was a case where a variance was being offered to U.S. Steel at the Edgar Thomson works, entirely within the boundaries of what was expected to be a successful regulatory process. What would happen years down the road, after U.S. Steel began challenging the entire process in court, or when the Variance Board found that it had to deny rather than extend variances to recalcitrant polluters?[67]

For many Pittsburghers the threat of job loss trumped environmental concerns, making the language of layoffs an increasingly powerful rhetorical tool as the economy worsened. This resulted in more serious blowback against GASP action. For example, James Mawhinney wrote into the *Post-Gazette* in complaint, arguing that GASP "has practically destroyed their credibility in the last several months with their irresponsible statements" and urging GASP members to "reexamine their leadership [and] . . . amend their by-laws to require that a quorum for any meeting include at least two persons who are unemployed steelworkers. Perhaps then their judgments will be tempered with the reason necessary" for the economic situation. In response to this newly critical tone, GASP altered its own rhetoric. The group's validity claims highlighting children's well-being, public health, and the conservation of environmental resources for future generations were almost entirely replaced in the debate over J&L and U.S. Steel by language concerning corporate irresponsibility and the importance of the rule of law to civilized society.[68]

Once it became clear that the largest steelmakers in the region were not going to willingly comply with air pollution control laws, the worst-case scenario of complete plant closure—which no regulator or environmentalist ever actually advocated—became a part of the debate over local air pollution control. In 1971, the *Post-Gazette* noted that the Clairton Coke Works "employs 4,500 persons but approximately 300,000 persons are economically dependent on the plant, and 30,000 jobs are directly or indirectly involved." While no one was advocating the closure of the works, the *Post-Gazette* reported that "the Appeals Board said U.S. Steel holds the 30,000 jobs over air pollution authorities here 'as a kind of blackmail.'" In 1973, during deliberations concerning U.S. Steel's activities at Clairton, Judge Silvestri brought up the apocalyptic, worst-case scenario in a discussion with the *Post-Gazette*, estimating that "more than a million workers would be affected directly and indirectly if the Clairton Works—which he described as the largest plant of its kind in the world—were to close." Even beyond the specific example of Clairton, the specter of job loss was described in all components of local industry: local papers carried threats from the United Mine Workers president that "Coal miners don't want to be pitted against environmentalists in a battle between jobs and clean air, but that's what present government policy ensures."[69]

In 1974, GASP president Arthur R. Gorr exchanged letters with the president of USWA Local 1843 on the subject of jobs and the environment, following a *Press* article about possible job loss due to environmental pressure. "Concern for the preservation of jobs has been a primary consideration in the activities GASP has undertaken to achieve a cleaner environment," wrote Gorr reassuringly, before becoming more explicit:

> *As a matter of fact, as I think back over the five years of existence with GASP, I am impressed with how patient we have been with companies like U.S. Steel and Jones & Laughlin. Frankly, I am of the opinion that if we had not been concerned about your jobs, we would have by now used every legal weapon to shut down [the polluting companies].*

Gorr continued by arguing that "job security blackmail has been one of the principal weapons used by heavy industry, steel in particular, to avoid spending money on pollution control." Whether or not the job threat was blackmail, it was certainly effective. By 1975 Madoff herself began to employ a pollution-for-jobs quid pro quo: "Before I vote to modify air pollution regulations," she said when a request for changes came before the advisory committee, "I want to see a definite commitment from J&L that the firm is going to remain in Pittsburgh."[70]

As the American economy—and the steel industry's economic health—worsened throughout the 1970s, GASP was forced to respond to a broader array of economic ills. Gorr argued in the written draft of a public speech that:

We have noted in many discussions of our current inflation, the costs of environmental or antipollution controls are often blamed by businessmen as a cause of higher costs. . . . I hope you will keep in mind, as some industrialists try to use inflation as an excuse for weakening environmental laws, that we can expect the cost of living to really soar if we fail to rescue and preserve our most fundamental resources: air, water, and land.[71]

The job loss card was played most explicitly in 1976, when U.S. Steel took out full-page ads in local newspapers to declare that unless it received unspecified local support, it could not continue operations in the region. "Did we make a mistake by locating our new headquarters building in Pittsburgh?" asked the ad in enormous font, covering nearly a third of page. U.S. Steel chairman Edgar Speer's image appeared below the incendiary query, and the text below laid out U.S. Steel's economic contributions to the region, including forty thousand jobs and $20 million in taxes. The text continued, wondering if, when U.S. Steel evaluated where to build and operate in the future, "will the governmental and environmental climate favor selection of the Pittsburgh area?" The ads so infuriated Madoff that she countered, borrowing $4,000 from an anonymous donor to respond with a full-page Enviro-SOS ad of her own, arguing that "the extensive advertising campaign by U.S. Steel was a well-orchestrated attempt to intimidate our community through veiled threats of economic blackmail."[72]

Despite GASP's best efforts, the rhetoric of jobs versus the environment became stronger as the plight of the declining steel industry became more obvious. Speer was a skilled practitioner of this rhetoric, arguing in 1977 that pollution control measures were crippling U.S. Steel's economic health: "Ultimately that will reduce US Steel's position in the Mon valley. It has to." Local papers translated this threat for their readers: "Speer had a tough message today for the Monongahela Valley. The message: Ease off environmental regulations or we'll pull out." Speer's statements on Pittsburgh jobs intensified the local debate through the late 1970s. Clairton's city council passed a resolution in 1979 "supporting the position of the steel industry and the United States Steel Corporation on the issue of unreasonable environmental standards which would affect the city of Clairton."[73]

County residents were very receptive to this argument: in 1976 public hearings, a group known as COKE (Clairton Organizes to Keep Employment) argued that they were committed to defending the industrial jobs their town depended on, and were "greatly concerned that the recent proposed consent agreement relative to the [coke works] will eventually lead to economic chaos in Clairton." There was extensive debate, never fully settled, about whether COKE was genuinely grassroots or was a front for steel producers. USWA local official Danniel Hannan, no stranger to controversy, declared that "U.S. Steel was COKE's grandfather," that the members were "pawns of U.S. Steel," and that they represented "a diabolical attempt to take over . . . through subterfuge and blackmail." But COKE members declared that they were entirely genuine defenders of local jobs: "We decided we'd better do something," declared COKE founder and McKeesport small-business owner Priscilla McFadden to the *Press*. "We worked until 2 in the morning—days, nights and weekends. I don't like anyone to imply to the contrary."[74] At about the same time as COKE's arrival, SING (or Survival with Industry Now through Growth) directly attacked and ridiculed GASP in the press, calling GASP leaders "a convention of kooks-of-the-world-unite in nurseryland" and accusing them of pressuring state attorneys into the attempt to remove Silvestri from the Clairton case. SING, if it existed at all, did not leave any mark beyond the occasional letter to the editor. But SING's supporters actively exploited jobs versus environment rhetoric. "I believe 'GASP' should spend more effort and time in trying to bring jobs and industry to Pittsburgh instead of being a destructive group driving industry out of Pittsburgh," wrote one. While it is possible that SING and COKE did receive corporate support, and while it is significant that these groups disappeared just as soon as U.S. Steel began decreasing operations in some regions, it is still quite possible that these groups were a genuine grassroots reaction to economic fears in a company town. There is no definitive answer about the provenance of these groups; about the only thing that is clear is the enduring pattern of acronym-named, grassroots organizations using the language of citizenship.[75]

GASP was fully aware of the power of the jobs versus environment rhetoric, but before the economic downturn of the mid-1970s, the organization felt that the jobs argument could be overcome. In 1972, GASP's third president, Ann Cardinal, told a journalist of an aphorism she'd learned from a nun in Braddock: "When there's dust on the windowsill, there's bread on the table." It was a more refined way of saying that "Smoke means jobs," but in referencing bread on the kitchen table, the saying literally hit home. In a story headlined "Citizens Wising Up," Cardinal went on to say that citizens were becoming more resistant to that argument: "They're not accepting those statements anymore." In March 1972 Pat Newman and Michelle Madoff combined to write an op-ed for the *Press*, in response to a syndicated op-ed by North American Rockwell CEO William F. Rockwell that had decried the high economic costs of pollution control: "Without a healthy ecology we will have no economic health," replied Newman and Madoff. "The very survival of business depends on the survival of air, water, and land."[76]

But no matter how well responses were phrased, it seemed difficult for GASP to counter the threat of job loss in the midst of a looming economic downturn, so it might make sense that the most-used rhetoric of the group in the J&L and U.S. Steel cases came to center around corporate responsibility and the rule of law. In fact, this strategy of replacing the language of women's care for future generations, so prominent in the group's educational and fund-raising efforts, with the language of public involvement, participatory democracy, and the rule of law was clearly most effective in promoting public involvement in the enforcement process.[77]

As in testimony before the Variance Board, in this public debate the attitude displayed by the corporation was just as important as the facts of the case. GASP never missed an opportunity to describe the large corporation as being insensitive to citizens' democratic wishes, often describing "U.S. Steel's public-be-damned attitude." "Their contempt," journalist Roy McHugh recalled Pauline Nixon saying, "is incredible." Never at a loss for words, Michelle Madoff called the company "a retarded giant" and a "gigantic corporate delinquent." But GASP's best rhetorical advantage lay in the fact that both J&L and U.S. Steel were violating existing law, allowing GASP to argue *for* the rule of law rather than *against* corporations in a letter to the county commissioners: "GASP feels this suit is a key example, an important first step in leading our county to new respect for law and order, freedom from the economic handicap of rampant industrial pollution, and renewed public faith in our public officials." GASP sent the same letter, urging the "strict enforcement of the law" in order to avoid "disrespect for the law," to Governor Milton Shapp.[78]

While it makes sense that GASP adopted "the rule of law" and "renewed public faith in our public officials" as a rhetorical trope in the same year that Watergate was becoming a household word, the use of the language of "law and order" might also explain GASP's deteriorating relationship with U.S. Steel. When GASP and U.S. Steel both appeared to be "working within the system" in an honorable manner to find a negotiated settlement, GASP was prepared to defend U.S. Steel, and the county's system of air pollution control, against the EPA's unwanted intrusion. But when U.S. Steel reneged upon its own consent decree, the corporation appeared to be mocking GASP's commitment to finding consensus with industry through the mechanism of the Variance Board. The relationship quickly turned adversarial, and GASP's language became more critical of big industry.

State and local regulators began to pick up this language after U.S. Steel proved resistant to regulation. When the state DER announced that it was rejecting the U.S. Steel cleanup plan, its legal counsel provided an explication of what GASP called U.S. Steel's "public-be-damned" attitude: "[Marvin] Fein said the company has shown a complete lack of good faith in the cleanup effort of Clairton and accused the steel firm of a callous disregard for the health and welfare of Clairton area residents." Lack of good faith in negotiations spoiled the atmosphere of cooperation that local politicians, eager to work out a voluntary consent decree, had previously celebrated.[79]

None of the rhetorical tools at GASP's disposal—citizenship, participatory democracy, maternalism, expertise, or the rule of law—was sufficient to best the threats of job loss and the worsening economic situation. In 1976, U.S. Steel ended its negotiations behind closed doors, excluding the public. Both J&L and U.S. Steel sidestepped the enforcement of county air pollution regulation by elevating their cases to the courts, and repeatedly invoking their regional economic power. J&L's threat of closure forced GASP to modify its rhetoric to include both jobs and environment, while U.S. Steel took its argument to the newspapers, threatening the unthinkable prospect of pulling out of the region in response to county pollution control. Industry's reactions and economic concerns had severely blunted the impact of environmental reforms in the region, and damaged hope of united progress through a shared system of regulation, implementation, and enforcement. While the variance process continued, the promise of its first years was tempered by reality in the mid-1970s.

• • •

The impact of new interpretations of citizen standing on the environmental movement in the United States is hard to overstate; in the case of GASP, it defined the group's identity and range of action. From an organization that described itself as "work[ing] within the system in a responsible manner" to achieve environmental goals, GASP's use of citizen standing rights in the legal battles with Jones & Laughlin and U.S. Steel transformed the organization into a litigious opponent of major corporations. The slow process of enforcing local law altered the group's rhetoric and goals as well as its understanding of the relationships between government, citizens, and regulated industries. The end result was an organization much more intensely critical of industrial corporations, its leadership somewhat radicalized by the experience of dealing with recalcitrant polluters, its range of action much more likely to include litigation not as a last resort after the immensely preferable negotiated agreement, but as an unavoidably necessary mechanism in the policy process.[80]

After its experiences with J&L and U.S. Steel throughout the first half of the 1970s, GASP became an active intervenor in major legal actions throughout the following decades. In the case of U.S. Steel, the failed consent decree of 1972 was followed by the entirely new state-county consent decree approved by the EPA in 1976. That decree was superseded as a result of a lawsuit filed in 1979 by the EPA, GASP, the commonwealth and county, and Local 1397 of the USWA. That suit forced U.S. Steel into signing yet another replacement consent decree on the subject of Clairton, this time directly with the United States, on July 10, 1979. GASP, represented in part by longtime member Edward Gerjuoy, was an intervenor in all levels of this case, granted standing in the case by federal district court. When U.S. Steel challenged GASP's right to intervene in 1980, that court responded by defending both the idea of public involvement and the specific contributions of GASP. "[W]e believe GASP has a right to be involved in these proceedings" wrote the court: "Congress explicitly recognized the importance of citizen participation; moreover, we expect that GASP's role may be helpful to the court in an area in which most federal courts lack scientific expertise." Even with GASP's involvement, enforcement of the many versions of the consent decrees dragged out through the 1980s and 1990s. As the steel industry continued to crumble, plant closures soon overtook attempts to regulate pollutants.[81]

Developments at Jones & Laughlin after 1975 followed the same general pattern as events at U.S. Steel. The negotiated consent decree of 1975 held for J&L, while local citizens groups continued to press for enforcement. GASP kept its own 1975 suit active against J&L for six years, so that it could be a party to all court discussions of the negotiated consent decree. Even after J&L was bought out and became part of LTV Steel, GASP continued to press for intervenor status in a variety of lawsuits, demanding payment for past civil penalties before the company's Hazelwood sites closed completely in 1998. GASP argued to the court that as a citizen intervenor,

> *GASP desires to be involved in any attempts by the parties to determine the terms and conditions of a possible settlement or order, to participate in any discovery that may be commenced by either party and to initiate its own discovery where appropriate, and to participate—through offering testimony and other evidence and examining and cross-examining witnesses and offering oral and written arguments to the Court—as part of any hearing or trial.*

This is the very definition of "substantive" public involvement that the citizen environmentalists wanted. In pursuing citizen standing in the many legal proceedings related to Pittsburgh's air quality, GASP became an active part of the process, with access to information, the right to appear in court and the power to bring suit against corporations and governments. Even after LTV closed its Hazelwood coke plants and entered bankruptcy, GASP maintained an active role in negotiations related to past civil penalties and whether the bankrupt LTV would be forced to pay them.[82]

In addition to its continued involvement in the enforcement of air pollution law in regard to J&L and U.S. Steel, GASP continued to use its standing in the court to intervene in a variety of suits not specifically related to the two companies. For example, GASP brought suit against the EPA in 1978 to force action regarding air pollution standards for basic oxygen process furnaces, a tool necessary for some types of steelmaking. When that case worked its way to the United States Court of Appeals, GASP had been joined by the National Resources Defense Council and Friends of the Earth, while the American Iron and Steel Institute had intervened on the side of the EPA.[83] In 2002 the Sierra Club and GASP combined to bring suit against then-EPA administrator Christine Todd Whitman, to force the EPA to make a final decision concerning attainment of air quality in some specific regions, as allowed under federal law. Thirty-two years after it was enshrined in the Clean Air Act, the rights of the citizen to standing in court continued to allow the substantive inclusion of citizens groups in air pollution policymaking.[84]

The importance of standing in enabling public involvement and the eventual transformation of air pollution control policy in Pittsburgh points to the significance of forces other than the national evolution of environmental philosophy as formative in the modern environmental movement. The transformation of air pollution policy in Pittsburgh had less to do with a national environmental movement, and more to do with a fundamental reorganization of the relationship between citizen and state, a part of the wide-reaching social revolutions of the 1960s.

The changing definition of citizen standing empowered not only local environmentalism, but also national environmental groups and the EPA itself. As one observer put it, environmental litigation, rather than being a sign of conflict, has become a mark of cooperative problem solving:

> *For participants in the environmental regulatory process, litigation even offers certain advantages for cooperative behavior. Litigation is not viewed as a last-resort strategy reserved for outsiders, as it is ordinarily thought to be, but rather as a legitimate institutional process for carrying on business as usual.*[85]

Environmental groups without standing, no matter how well organized or persuasive, would have been entirely excluded from a crucial component of policymaking and enforcement.

The Clairton saga is not yet completed, and its current cycle echoes previous installments with bitter irony. In 2007, the Allegheny County Health Department announced that it had signed yet another consent decree with U.S. Steel to clean up the Clairton Coke Works, which included a $396,000 civil penalty "to settle past violations" for which the company had never paid in full. And in response to a revitalized steel industry in 2008, U.S. Steel applied for a permit to greatly expand coke production at the Clairton works, promising to spend more than a billion dollars to modernize the production process. This was the very modernization that the citizens, city, county, state, and federal authorities had urged for the previous decades. After much debate over the merits of increased production, no matter how well regulated, permission was eventually given. But in a cruel twist of fate, the Clairton plan was put on hold in the economic downturn of 2009. Incredibly, as the fortieth anniversary of both GASP and the modern regime of county air pollution regulation took place in the fall of 2009, the U.S. Steel Clairton saga continued. U.S. Steel and regulators continue an uneasy dance, the facility still drags down air quality in the region, and the new county executive is exploring the idea of eliminating or streamlining local control of air pollution in order to encourage economic development.[86]

Conclusion

THE SIGNIFICANCE OF THE CITIZEN ENVIRONMENTALISTS FOR THE MODERN ENVIRONMENTAL MOVEMENT, IN PITTSBURGH'S EXPERIMENT, AND IN GASP

The citizen environmentalists were created by a confluence of forces, from the opportunities for public involvement in federal legislation, to the legal philosophy of citizen standing, to the late-1960s support for the idea of participatory democracy. In turn, these activists used the most powerful tools available to them to push their way into the environmental policymaking process. They used not only the emerging ideology of environmentalism but also the language of citizenship, the rhetoric of maternalism, and persuasive claims of scientific expertise. The demographics of their membership connected them to the dense network of women's civic organizations and the professional authority of men's academic and technical institutions. They used these tools and capacities to substantively change policy institutions and outcomes. But what, exactly, is the significance of the rise of citizen environmentalism? How can we assess their impact? In short, as I often ask myself and my students, *so what?* There are three major components to the significance of the citizen environmentalists, including a new understanding of modern environmentalism, the considerably improved environment of Allegheny County, and the persistence and legacy of GASP.

• • •

There is no doubt among historians of the United States that the last half of the twentieth century witnessed an unprecedented growth in environmental concern. Historians have already presented arguments, quite convincingly, about *why* and *how* this occurred. Samuel P. Hays has written that Americans newly free of the economic and military demands of the Great Depression and World War II turned their attention to their own long-delayed prosperity, health, and enjoyment, creating a broad-based, popular movement for the maintenance and enjoyment of high standards of living. Other historians have outlined the emergent philosophies of environmentalism and the political mechanisms that transformed this environmental consciousness into legislative reality. Still others point to the rise in decidedly unnatural and clearly polluting postwar technologies such as the chemical industry, plastics production, and nuclear arms, and the public aversion to those innovations.[1] But while the *why* and *how* have already been discussed, the *significance* of the modern environmental movement has not been sufficiently addressed by historians. Beyond the obvious introduction of a new area of political dispute and the popularization of a new philosophy, what is the historical meaning of the rise of environmental thought and action in the last half of the twentieth century?

The answer lies not within the environmental movement itself, but rather in the movement's relationship with contemporaneous events. While all of the concerns of environmental reform had previously appeared in American history—public health, the decay of the city, conservation of wildlands, the alteration of landscape for agricultural purposes, the maintenance or preservation of flora and fauna—what was different about the modern environmental movement was a fundamental change in the relationship between citizens and their government.[2] What was different in the 1960s and 1970s was that individuals could use the power of mass media, the tactics of mass protest, the philosophies of individual rights, and a popular belief in individual freedom to agitate for broad social change. In short, the historical significance of the modern environmental movement has less to do with the *environment*, and more to do with the *movement*. The environmental movement was a part of a rights revolution, a fundamental redistribution of power from a previously dominant, corporatist governing consensus to a pluralistic, diverse, contentious, and fractious public.

The civil rights, women's liberation, and environmental movements each waged wars to empower the previously excluded individuals and groups they represented. Their battles quite often took place within courtrooms and judicial chambers and revolved around the rights of individuals to take part in a free and open society. The sit-ins, protests, marches, consciousness raising, lobbying, letter writing and litigating of these movements represented a renovation of the public sphere, forcefully redirecting the nation's attention to social matters previously excluded from meaningful debate.

The same might be said of Pittsburgh. It had been known as the Smoky City for more than a century, but it was only in concert with a nationwide revolution in individual rights that environmental concerns in Pittsburgh could be translated into political pressure or legal rights. The fundamental transformation of Pittsburgh's local air pollution control mechanisms occurred at precisely the same time that citizens were redefining their individual rights in relation to government nationwide. GASP members challenged the narrow economic definition of "direct interest" to intervene in air pollution control matters on the county level in 1969. GASP, and other citizens groups around the nation, took the opportunity offered by the federal government to become involved in the creation of new state implementations of the federal Clean Air Acts. This surge in public involvement in policymaking and enforcement occurred simultaneously with a revolution in the ability of citizens to bring lawsuits against industries, corporations, and regulators themselves. Once they had argued their way into a position of influence, they lobbied to create permanent positions for public involvement in the relevant regulatory and enforcement bodies formed in 1970.

The regulatory world left by the green revolution was fundamentally transformed in its approach to citizen participation. Policy historian Richard N. L. Andrews notes that "[s]ince 1970 there has been an unprecedented increase in the number and effectiveness of citizen organizations advocating environmental protection." Environmental policy scholar Richard Munton writes that "[o]utside the corporate domain it is, today, quite difficult to find examples of environmental decision-making where there has been no public consultation or other form of public involvement in the process." In the United States, no federal or state construction project of any size can begin without a lengthy environmental impact statement heavily dependent upon the involvement of the affected citizenry. Few national or state regulations, whether related to environmental matters or not, may likewise go into effect without some sort of public comment process, as the model of public presentations, comments, and revisions outlined in the Environmental Impact Statement process has become institutionalized and spread to several different fields. The end result is a legalistic, bureaucratic, and technocratic decision-making regime; but also one that must, in some manner, explicitly include the public in the process. The dividing point between the corporatist and pluralistic regimes is 1970, as historian Frank Uekoetter and legal scholar Noga Morag-Levine both point out, precisely central to the surge in environmental activism.[3]

This phenomenon may, in fact, be the reverse of the dire predictions of Robert Putnam. While the Americans portrayed in his influential work spend less time in groups building the social capital essential for political organizing, they have greater access to a governing process more inclined to pay attention to their concerns and individual rights. Rather than a withering of civil society in the late twentieth century, Americans have witnessed a great flowering of their influence, coupled with legislation that not only encourages but also is dependent upon public involvement. The late twentieth century surge in environmental organizing complicates Putnam's declensionist narrative. There may indeed be ways in which civil society in the United States regressed, but that decline took place in the same time period in which civil rights organizations, women's groups, environmental activists, and others gained access to arenas of public debate, political representation, and legal defense. For these groups and others, the period of the 1960s and 1970s was a time when both federal and individual powers were strengthened, redefining citizenship to include the marginal, redefining legal rights to include the public interest, and redefining federal power to include intervention in local matters. In these instances, federal intervention and individual rights are both mutually dependent and dialectically regenerative: each makes the other possible.

• • •

Beyond the historiographical argument, how can we assess the impact of the citizen environmentalists in Pittsburgh? This is a surprisingly difficult question to answer, as the undeniable successes of pollution control get mixed in with the effects of deindustrialization, demographic transition, and the persistence of particularly troublesome pollution sources. Pittsburgh is a very different place today, and its transformation has multiple causes.

In Allegheny County, the experiment begun in 1969 was relatively successful for many of the actors involved. Politicians responded to public outcry by creating a comparably more stringent air pollution control regime; a flexible and independent Variance Board allowed for moderate but progressive implementation; systems of structured public involvement and an involved and expert public created political cover and oversight; and a local Bureau of Air Pollution Control was empowered to steer industry into compliance with regulations. The regulatory regime begun under Article XVII allowed Allegheny County to respond to the unique challenges of the Smoky City, in a way that regulation from Harrisburg or Washington could never have done. The region's unique geography, complex local politics, collections of technical expertise, and distinctive industrial history demanded the creation of a new and experimental regulatory framework, one that was largely successful in pushing many polluters into compliance with local control.

In Pittsburgh, the experiment in harnessing the public sphere as a means of supporting state regulatory power also had very real success. The many institutions of civil society—the educational institutions, older conservation groups, churches, women's clubs, unions, newspapers, researchers, activists, medical associations, neighborhood groups, and new environmental organizations—jostled against each other as they alternately pressured and defended regulators and the corporations they regulated. In the quasi-state public space of the Variance Board, and in the pages of the local newspapers, they debated competing values and views of the ideal city with representatives of the government and of corporations. As a measurement of the power and healthiness of civil society, this study indicates that Pittsburgh possessed a particularly lively specimen, and one that could focus both political pressure and considerable local expertise on the problem at hand. As an ironic result of the concentration of industry and industrial pollution in the region, Pittsburgh had a concomitant concentration of the exact type of technical knowledge and professional expertise necessary to have representatives of the public cogently engage in difficult technical debates. Interestingly, the increase in an actively involved citizenry appeared to be a correction, a pendulum swing back from previously unchallenged or unfettered corporate power.

If there's a simple moral to this part of the story, it can be summarized by the phrase "If you build it, they will come." The dialectic relationship between the federal enactment of public-hearing legislation and the genesis of citizens' activist organization can be expressed almost as plainly: if an opportunity for substantive public involvement is created, a substantively involved public will emerge. The opposite is also true: limiting opportunities for citizen involvement, either through a reversal of citizen-standing suits or through the elimination of federal oversight of local environmental practices, will likely disengage that very same public.

That having been said, the inclusion of the public and experimentation with a new regulatory regime ran into a considerable challenge when it came to large-scale industry and the threat of job loss. Even GASP knew the score in 1975, as recorded in the conclusion of its film *I Belong Here!* After the narrator asks about the results of the new push for regulation, a turtle appears on the screen as an unnamed GASP volunteer responds: "The results? It seems that the major breakthroughs are always just beyond our grasp. Steel's promises, good intentions, signatures on consent decrees, have resulted in very little except protracted court cases. And the pollution continues." Extensive legal machinations on the part of some put limits on the success of this new method of policy implementation.[4]

There would be greater success after 1976 in regulating Clairton and other large industrial sources of air pollution nationwide. Armed with the technical expertise, legal background, and hard-won experience gained in the first half of the decade, regulators, experts, and activists alike elevated their debates to a national level, resulting in more wide-ranging federal consent decrees on the subject of Clairton in 1979 and substantive changes in 1991. These efforts resulted in measurable improvement in the region's air quality by the late 1970s, with the county declaring in 1978 that particulates and sulfur dioxide had decreased by 17 and 38 percent respectively in the previous four years, improvements that continued over time, though not as drastically. On a larger time scale, and counting the escalation to the federal level, there is some considerable success in regulating major industrial sources.[5]

However, the entire point of the passage of Article XVII in 1969 was to keep control of the regulatory process local while responding to increased public pressure with real or perceived progress. It surely could not have been the intent of the county commissioners that the process of regulating Clairton would only be successful with a move to the federal level that involved a small army of lawyers and agencies, and which could not be considered even partially complete until decades after the passage of Article XVII. Neither could GASP, with its emphasis on assisting the public in gaining substantive access to policymaking institutions, have intended that the process of local control take a detour to Judge Ellenbogen's closed-door negotiations in 1976. Those back-room agreements marked a momentary return to the consensus-oriented policy process of Mayor Lawrence and the ACCD, not the pluralistic debate of the Variance Board.

As an extension of this point, it would be difficult indeed to argue that the environmental movement pressured the steel industry out of existence. Disturbingly, it was largely deindustrialization, globalization, and large-scale economic trends that made Pittsburgh a much cleaner place. Local air pollution control laws had limited success in the regulation of large polluters, and were it not for deindustrialization and the extensive idling and closure of plants in the early 1980s, intransigent industry would have continued to bedevil citizen environmentalists.

In fact, economic forces shuttered many of the major sources of industrial pollution, regardless of the plans of environmentalists or regulators. These closures were far beyond the power of local government either to cause or to prevent. U.S. Steel's Homestead Works was closed in 1984 due to economic conditions, not actions taken by the Allegheny County Department of Health, and the historic factory site was ignominiously replaced by an open-air shopping mall in 1999. After decades of financial decline, LTV Steel filed for bankruptcy in 2000. What was once J&L's mighty Southside plant has since been replaced by a development featuring an REI, a movie theater, a Steelers' practice field, and a surprisingly good Irish pub. J&L's former Second Avenue location, across the river, hosts a prosperous business and technology park. J&L's Hazelwood coke plant closed in the 1990s, and the land sits abandoned. More recently, U.S. Steel's promise to invest $1.1 billion in a state-of-the-art Clairton Coke Works, planned in 2007 and derailed in 2009, was not prompted or prevented by local regulation or environmental politics. Rather, the plan to reinvest in advanced environmental controls for the coking process, for decades the dream of Pittsburgh environmentalists, was proposed by the company on its own due to improving prices for steel and then cancelled due to the national economic downturn. No, Pittsburgh is appreciably greener today at least partly by historical trends; and while that observation exonerates environmentalists from blame for industrial downturn in Pittsburgh, it demonstrates the limited capacity of local government and environmentalists to resist economically appealing but environmentally disastrous industries in the future.[6]

Indeed, recent stories in Allegheny County reflect this argument. After decades of attempts to regulate Clairton and ameliorate its environmental impact, new findings indicate that residents of the Clairton region are still twenty times more likely to contract cancer than other Americans. Residents of Clairton and its neighboring community Glassport can, if they desire, boast the third and fourth highest risks of cancer from airborne toxics in the nation. Though the residents of the newly green Pittsburgh aren't pleased with the news, the Clairton site continues to drag down the region in national rankings of air quality. Despite these facts, conversations continue about eliminating or restricting local control of air pollution in order to attract and promote industrial development. Economic concerns have incredible weight on policymaking, circumscribing even the remarkable success of Pittsburgh's experiment.[7]

The constraints on Pittsburgh's regulators and environmentalists might itself stand in for the limitations of the modern environmental movement. At the start of the twenty-first century, there has been a critical reappraisal of the goals and accomplishments of the environmental movement, as embodied in the work of Michael Shellenberger and Ted Nordhaus. The bare fact is that the staggeringly ambitious goals of the environmental movement remain elusive for a limited coalition of special-interest groups outside of the political parties whose goals often run counter to short-term profits and tax revenues.[8]

The most powerful tool of the environmentalists described in this book has been called into doubt. Citizen standing has been challenged in courts across the nation. It is ironic that the active rollback of citizen standing as a judicial philosophy comes just as historians are beginning to understand the extensive policy impact of pluralistic regimes.[9] This pressure comes just as convincing case studies from policy experts promote the utility of organized, structured public involvement: "Local knowledge is essential," concluded the head of a panel organized in 2008 by the National Research Council to assess the importance of public participation in environmental policymaking, and most other studies tend to similar conclusions about the utility of structured public involvement. It would be strange indeed if a backlash against citizen involvement left a short, twenty-five-year period at the end of the twentieth century as the high-water mark of public involvement in environmental policy.[10]

• • •

And what about GASP? Whatever ironies historians might uncover, the process of local air pollution control goes on in Pittsburgh. Forty years after it was founded, GASP continues to be active in local and national environmental politics. It deals with many of the same actors, as well. Many of the original activists, lifelong Pittsburghers, remain a central part of the organization. Joined by new generations of environmental activists, many of whom came to Pittsburgh after the last steel mill in the "Smoky City" closed, GASP's members still organize in much the same way. They raise funds, lobby for policy change, and show up at what are often sparsely attended and tedious hearings. They often use similar rhetoric, as well: the recent triumph of the group, a campaign to retrofit emission controls on diesel school buses, used arguments such as "Diesel pollution has a profound impact on children's health," and "Healthy buses = healthy kids." Maternalist language once again allows the group entrée into the otherwise contentious issue of regulating auto emissions, a new frontier for the organization. Along with claims of citizenship, the rule of law, and scientific expertise, these arguments are used to marshal support for air pollution control, and have been successfully used for forty years.[11]

In fact, for decades now, simply being present was one of the most significant accomplishments of the citizen environmentalists. Even when agency hearings were opened to the public, the opportunity costs in time and energy still made it unlikely that the public would actually attend. As a result, GASP was often the only representative at public hearings in Pittsburgh, as the local press noted: "A public hearing . . . drew a grand total of one outside participant yesterday. . . . [A] board member of [GASP] was the lone member of the public to show up." By dint of showing up, year after year, members of the public could intervene with surprising success. GASP members not only showed up, they encouraged others across the nation to do so, promoting the ideal of an engaged citizenry in any way possible.[12]

GASP members, and the many other Pittsburghers who became involved in environmental activism in the 1970s, had a lasting effect both on the city and on the wider nation. In countless instances, Pittsburghers involved in the air pollution debate of the early 1970s went on to state or national positions, taking hard-won experience and knowledge with them. Michelle Madoff, of course, went on to the state Citizens' Advisory Board as well as the city council, while Robert Broughton became chair of the state Environmental Quality Board. Anthony Picadio, in addition to his stint as the second chair of the Variance Board, went on to serve as Pennsylvania assistant attorney general with a special brief in environmental enforcement matters. Morton Corn was already a nationally known expert before contributing to GASP, but his leadership of OSHA, support for an important coke oven emissions standard in October 1976, and subsequent academic career at the Johns Hopkins Bloomberg School of Public Health vaulted him to a wider stage. Lester Lave is still a nationally known expert in economics, air pollution, and regulation, and has worked extensively with the EPA and private industry. Ann Cardinal moved from local environmental activism to working as a spokesperson for the EPA Region III in the 1980s. Gerald P. Dodson, the county solicitor closely wrapped up in the Clairton saga, went on to serve as chief counsel for the Health and Environment Subcommittee in the U.S. House of Representatives, and with the Solicitor's Office in the U.S. Department of the Interior, arguing before the Supreme Court in multiple cases. He has since become an intellectual property lawyer specializing in high-tech and scientific issues. Bernard Bloom took his knowledge to the EPA as an iron and steel mill emissions control expert, before serving as director of a county air program in Maryland and being employed as a nationally recognized consultant in indoor air quality. Even Charles O. Jones, the author of an early study of the Pittsburgh air pollution debate, went on to become the president of the American Political Science Association and a senior fellow at the Brookings Institution. Those who experienced the air pollution debate in Pittsburgh spread that knowledge far and wide.

However, forty years after the contentious outpouring of public anger in the fall of 1969, local control of air pollution control in Allegheny County is still a controversial undertaking. With the creation of a new county executive's office and conspicuous decline both of the population of Pittsburgh and the size of local government, pressure is building to streamline local regulation in order to woo new employers. While a transformed region may indicate the need for a new regulatory regime, it would be unwise to throw out the baby with the bathwater. The fact that Pittsburgh has local air pollution regulation and the inclusion of structured public involvement in policymaking flows from the unique history and challenges of the region. Pittsburgh's air pollution regulation is different from the rest of the state because Pittsburgh's air pollution problem is different from the rest of the state. That local system has had functional and political success, even if the path to compliance was occasionally messy, convoluted, and litigious. Eliminating that system cannot guarantee increased economic development, but it can guarantee the sort of dissatisfaction, unrest, and anger demonstrated in the public hearings of the late 1960s and early 1970s, when citizens who felt themselves locked out of an essentially corporatist policymaking process waylaid unsuspecting elected officials.[13]

The example of GASP and other citizen environmentalists provides a blueprint for creating an active citizenry and fostering participatory democracy. If, as this study indicates, policy does indeed create participation, and participation is deemed an overall social good, then it is fairly simple to create participation: all one has to do is legislate it. This is an important conclusion, as it implies that similar results may be achieved in the future. As various theorists, observers and pundits argue that the health of a democracy may be measured by the amount of civic engagement of the populace, a remedy for lagging engagement is available: simply create opportunities for substantive and structured involvement in the process, and citizens, or at the very least those who claim to speak for citizens, will rise to claim those opportunities.

Empowered by federal requirements for public inclusion, inspired by countless contemporaneous social movements, and enabled by a changing legal definition of the rights of the individual citizen, community organizations embarked on a wide-ranging project to promote environmental thought, policy, and practice in the 1970s. This book argues that legal and political opportunities for citizen involvement created an engaged, active, and involved citizenry. The opposite is most likely also true: the absence of opportunities for meaningful participation might create an apathetic, disaffected, and powerless citizenry. And while government meetings may be more quiet and predictable without citizen input, they won't be the meetings of a democracy.

Notes

The bibliography includes a list of abbreviations used for archival sources.

Preface: The Citizen Environmentalists

1. The rhetorics of citizenship, maternalism, and scientific expertise may be conceived of as validity claims, as theorized by Jurgen Habermas in *The Theory of Communicative Action*, vol. 1, *Reason and the Rationalization of Society* (Boston: Beacon Press, 1985), 19.

2. Alice Felt Tyler, *Freedom's Ferment: Phases of American Social History to 1865* (Minneapolis: University of Minnesota Press, 1944); Elizabeth Clemens, *The People's Lobby: Organizational Innovation and the Rise of Interest Group Politics in the United States, 1890–1925* (Chicago: University of Chicago Press, 1997). For more on specific periods of public activism, see Gordon S. Wood, *The Creation of the American Republic, 1776–1787* (New York: W. W. Norton, 1972), 319; Wood, *The Radicalism of the American Revolution* (New York: Random House, 1991), 89–91; Robert H. Wiebe, *The Search for Order, 1877–1920* (New York: Hill and Wang, 1967). For an extended review of the literature discussed in this preface, see James L. Longhurst, "'Don't Hold Your Breath, Fight for It!' Women's Activism and Citizen Standing in Pittsburgh and the United States, 1965–1975" (Ph.D. dissertation, Carnegie Mellon University, 2004), 1–33.

3. John Dewey, *The Public and Its Problems* (New York: Holt, 1927); "At Clinton Dinner," *Washington Post* (January 20, 1995): A8.

4. Alexis de Tocqueville, *Democracy in America* (New York: Mentor, 1956), 206 and 208. For Tocqueville in the modern civic engagement debate, among many see Bob Edwards, Michael W. Foley, and Mario Diani, eds., *Beyond Tocqueville: Civil Society and the Social Capital Debate in Comparative Perspective* (Lebanon, NH: University Press of New England, 2001); Theda Skocpol, "The Tocqueville Problem: Civic Engagement in American Democracy," *Social Science History* 21 (1997): 455–79; Theda Skocpol and Morris P. Fiorina, eds., *Civic Engagement in American Democracy* (Washington, DC: Brookings Institution Press, 1999).

5. Edward C. Banfield and James Q. Wilson, *City Politics* (Cambridge, MA: Harvard University Press, 1967), 258.

6. Jurgen Habermas, *The Structural Transformation of the Public Sphere: An Inquiry into A Category of Bourgeois Society* (Cambridge, MA: MIT Press, 1989), 27; for an excellent application of Habermas on which this book is modeled, see Philip J. Ethington, *The Public City: The Political Construction of Urban Life in San Francisco, 1850–1900* (New York: Cambridge University Press, 1994).

7. James S. Coleman, "Social Capital in the Creation of Human Capital," *American Journal of Sociology* 94 (1988): 95–120; Robert D. Putnam, *Bowling Alone: The Collapse and Revival of American Community* (New York: Simon & Schuster, 2000), 19–26; Longhurst, "Don't Hold Your Breath," 19, nn. 21–23.

8. Robert A. Dahl, *Who Governs? Democracy and Power in an American City* (New Haven:

Yale University Press, 1961); Clarence N. Stone, *Regime Politics: Governing Atlanta, 1946–1988* (Lawrence: University Press of Kansas, 1989), 3–4. For citizen participation within regime theory, see Bernard H. Ross and Myron A. Levine, *Urban Politics: Power in Metropolitan America* (Itasca, Illinois: F.E. Peacock, 1996), 217–247. N.B. Barbara Ferman, *Challenging the Growth Machine: Neighborhood Politics in Chicago and Pittsburgh* (Lawrence: University Press of Kansas, 1996), 38–43 and 46–49; cf. Sherie Mershon, "Corporate Social Responsibility and Urban Revitalization: An Organizational History of the Allegheny Conference on Community Development, 1943–1990" (Ph.D. dissertation, Carnegie Mellon University, 2000), 153–221. On counterweights to state power, see Michael W. Foley and Bob Edwards, "The Paradox of Civil Society," *Journal of Democracy* 7 (1996): 38–52.

9. Gabriel A. Almond and Sidney Verba, *The Civic Culture: Political Attitudes and Democracy in Five Nations* (Boston: Little, Brown, 1963), 3–6. On the topic of civil society, among many, see Paul Wapner, "Politics Beyond the State: Environmental Activism and World Civic Politics," *World Politics* 47 (1995): 311–340; George Pettinico, "Civic Participation Is Alive and Well in Today's Environmental Groups," *Public Perspective* 7 (1996): 27–30; Rodger A. Payne, "Freedom and the Environment," *Journal of Democracy* 6 (1995): 41–55. For more, see Longhurst, "Don't Hold Your Breath," 20 n. 24.

10. Elizabeth S. Clemens, "Organizational Repertoires and Institutional Change: Women's Groups and the Transformation of American Politics, 1890–1920," in Skocpol and Fiorina, *Civic Engagement in American Democracy*, 81–110; Clemens, *The People's Lobby*; Almond and Verba, *The Civic Culture*, 151–154. On urban politics, see Susan S. Fainstein and Clifford Hirst, "Urban Social Movements," in *Theories of Urban Politics*, ed. David Judge, Gerry Stoker, and Harold Wolman (London: SAGE, 1995), 181–204; Vivien Lowndes, "Citizenship and Urban Politics," in Judge, Stoker, and Wolman, *Theories of Urban Politics*, 160–180; Ross and Levine, *Urban Politics*, 217–247.

11. For concise summaries of the field of environmental history describing the approaches of Hays, Gottlieb, Rome, and others, see J. R. McNeill, "The Nature of Environmental History: Observations on the Nature and Culture of Environmental History," *History and Theory* 42 (2003): 5–43; Char Miller and Hal Rothman, eds., *Out of the Woods: Essays in Environmental History* (Pittsburgh: University of Pittsburgh Press, 1997), xv; and Christopher C. Sellers, "Environmentalists by Nature: The Postwar America of Samuel Hays," *Reviews in American History* 28 (2000): 112–119.

12. Benjamin Kline, *First Along the River: A Brief History of the U.S. Environmental Movement* (San Francisco: Acadia Books, 2000), 80; Hal K. Rothman, *The Greening of a Nation? Environmentalism in the United States since 1945* (Fort Worth, TX: Harcourt Brace College, 1998), 120–121; Mark Dowie, *Losing Ground: American Environmentalism at the Close of the Twentieth Century* (Cambridge, MA: MIT Press, 1995); Philip Shabecoff, *A Fierce Green Fire: The American Environmental Movement* (New York: Hill and Wang, 1993); Riley E. Dunlap and Angela G. Mertig, eds., *American Environmentalism: The U.S. Environmental Movement, 1970–1990* (Philadelphia: Taylor & Francis, 1992); Deborah Lynn Guber, *The Grassroots of a Green Revolution: Polling America on the Environment* (Cambridge, MA: MIT Press, 2003).

13. For definitions of these groups, see Nicholas Freudenberg and Carol Steinsapir, "Not in Our Backyards: The Grassroots Environmental Movement," in Dunlap and Mertig, *American Environmentalism*, 27–37; Chapter 3 in this book.

14. William Cronon notes that one of "the greatest weaknesses of environmental history" is "its failure to probe below the level of the group to explore the implications of social divisions for environmental change." Cronon, "Modes of Prophecy and Production: Placing Nature in

History," *Journal of American History* 76 (1990): 1129; cf. Alan Taylor, "Unnatural Inequalities: Social and Environmental Histories," *Environmental History Review* 1 (1996): 6–19.

15. Theda Skocpol, *Diminished Democracy: From Membership to Management in American Civil Life* (Norman: University of Oklahoma Press, 2003), 136–138.

16. David Farber and Beth Bailey, eds., *America in the Seventies* (Lawrence: University Press of Kansas, 2004); Bruce Schulman, *The Seventies: The Great Shift in American Culture, Society, and Politics* (New York: Da Capo, 2002); Edward Berkowitz, *Something Happened: A Political and Cultural Overview of the Seventies* (New York: Columbia University Press, 2007).

17. Aldon D. Morris, *The Origins of the Civil Rights Movement: Black Communities Organizing for Change* (New York: Free Press, 1984), 282. Cf. Michael L. Clemons, "Political Mobilization in Sumter County, Alabama: Through the Prism of the Indigenous Perspective," *Journal of Black Studies* 28 (May 1998): 650–675; Rosemary D'Apolito, "The Activist Role of the Black Church: A Theoretical Analysis and an Empirical Investigation of One Contemporary Activist Black Church," *Journal of Black Studies* 31 (September 2000): 96–123. Periodizing the civil rights movement as a development of postwar America is a vast oversimplification; see the introductory chapter of Charles M. Payne and Adam Green, eds., *Time Longer than Rope: A Century of African American Activism, 1850–1950* (New York: New York University Press, 2003); Sean Dennis Cashman, *African-Americans and the Quest for Civil Rights, 1900–1990* (New York: New York University Press, 1993).

18. Belinda Robnett, *How Long? How Long? African-American Women in the Struggle for Civil Rights* (New York: Oxford University Press, 1997), 19; for more, see Steven F. Lawson, *Civil Rights Crossroads: Nation, Community, and the Black Freedom Struggle* (Lexington: University Press of Kentucky, 2003), 3–28; cf. Payne and Green, *Time Longer than Rope*, 2.

19. Morris quote from *Origins*, 288; Schudson quote from *The Good Citizen: A History of American Civic Life* (New York: Free Press, 1998), 255.

20. Flora Davis, *Moving the Mountain: The Women's Movement in America Since 1960* (Urbana: University of Illinois Press, 1999); Ruth Rosen, *The World Split Open: How the Modern Women's Movement Changed America* (New York: Viking, 2000). On the multiplicity of movements, see Myra Marx Ferree and Patricia Yancy Martin, eds., *Feminist Organizations: Harvest of the New Women's Movement* (Philadelphia: Temple University Press, 1995); Jo Freeman, "On the Origins of Social Movements" and "A Model for Analyzing the Strategic Options of Social Movement Organizations," in Freeman, ed., *Social Movements of the Sixties and Seventies* (New York: Longman, 1983), 8–30 and 193–210.

21. Quote from Estelle B. Freedman, *No Turning Back: The History of Feminism and the Future of Women* (New York: Ballantine, 2002), 87. For the many local movements, see Rachel Duplessis and Ann Snitow, *The Feminist Memoir Project: Voices from Women's Liberation* (New York: Three Rivers Press, 1998); Alice Echols and Ellen Willis, *Daring to Be Bad: Radical Feminism in America, 1967–1975* (Minneapolis: University of Minnesota Press, 1990); Vicki Ruiz and Ellen Carol DuBois, eds., *Unequal Sisters: A Multicultural Reader in U.S. Women's History* (New York: Routledge, 2000).

22. Adam W. Rome, "'Give Earth a Chance': The Environmental Movement and the Sixties," *Journal of American History* 90, 2 (2003): 527.

23. James T. Patterson, "The Rise of Rights and Rights Consciousness: 1930–1980," in Byron E. Shafer and Anthony J. Badger, eds., *Contesting Democracy: Substance and Structure in American Political History, 1775–2000* (Lawrence: University Press of Kansas, 2001), 201–224; Patterson, *Grand Expectations: The United States, 1945–1974* (Oxford: Oxford University Press,

1996), 452–453, 637–677; Berkowitz, *Something Happened*, 133–157; John D. Skrentny, *The Minority Rights Revolution* (Cambridge, MA: Belknap Press, 2002), 4–15.

Chapter 1: Power to the Public Hearing

1. "County Clean Air," *PPG* 9/25/69; "Potential Hazards," *PPG* 9/26/69; "County Weighs," *PPG* 9/27/69; "Clean Up County Air," *PP* 9/24/69; "Unclean Air Tragedy," *PP* 9/27/69; Charles O. Jones, *Clean Air: The Policies and Politics of Pollution Control* (Pittsburgh: University of Pittsburgh Press, 1975), 162–166.

2. "Clean Air Warriors," *PPG* 10/20/70; "'Breather's Lobby' Wins," *WSJ* 10/21/69; "Pittsburgh Cracks Down," *Business Week* (December 27, 1969): 66–67.

3. The Citizen's Clean Air Committee, from Dayton, Ohio, the Citizens Against Pollution from Portsmouth, Virginia, and the Citizen's Council for Clean Air from Indianapolis, Indiana, were only three of many. Fred M. Nathanson and Gwen Gwinn, "This Industry Cleaned Up the Air–With a Little Prodding," *American Lung Association Bulletin* (March 1975): 10–13; ff "Citizens Committee," Office of Air Programs Administrative Subject Files, 1966–1972, NARA RG 412. See the bibliography for a discussion of archival sources. The nearly eight hundred group names are from the National Center for Charitable Statistics database at http://nccs.urban.org/, Master-05/2006. For further discussion of these organization names see Chapter 3.

4. "A Citizen's View of Air Pollution," *PP* 9/26/69; "Citizens Demand," *PPG* 11/10/73; "U.S. Steel Forced," *NYT* 8/27/72.

5. General Promotional Brochure, "Don't Hold Your Breath . . . Fight For It!" "About GASP," all from CLP GASP. "Citizen's Advisory Council to the Department of Environmental Resources, 1973–1974 Annual Report," ACHD BAPC; ff "OC Citizen's Committee 1971 (1)," NARA RG 412; "Citizens Wising Up," *PP* 11/26/72.

6. Searches of 600, 585, and 542 U.S. newspapers in the 1950s, 1960s, and 1970s in the Access NewspaperARCHIVE database, http://www.newspaperarchive.com, viewed January 13, 2009. Results for the 1980s continue the trend, but come from only 213 newspapers.

7. These findings, and Chart 1.1, are from IRS 501(c)(3) data in Master-05/2006, NCCS.

8. Voice of America audio recording, *A Breath of Fresh Air*, CLP GASP.

9. "Stenographic Report . . . January 17, 1972," 264, ff "Implementation Plan Hearing, Jan. 17, 1972—Pittsburgh," PSA RG 43.36. See also hearing on December 3–4, 1971.

10. On the "rights revolution," see James T. Patterson, *Grand Expectations: The United States, 1945–1974* (New York: Oxford University Press, 1996), 453; John D. Skrentny, *The Minority Rights Revolution* (Cambridge, MA: Belknap, 2004); Michael Schudson, *The Good Citizen: A History of American Civic Life* (New York: Free Press, 1998), 242.

11. On definitions of "public involvement," see Frans H. J. M. Coenen, Dave Huitema, and Laurence J. O'Toole, eds., *Participation and the Quality of Environmental Decision Making* (Dordrecht: Kluwer Academic, 2000), 310. For the opposite of "substantive involvement," see George A. Gonzalez, *The Politics of Air Pollution: Urban Growth, Ecological Modernization, and Symbolic Inclusion* (Albany: SUNY Press, 2005), 14–16, 24–26.

12. Anne Payne, Jeff J. Shampo, and Eric C. Surette, "Parties," in *American Jurisprudence* 34, 2nd ed. (Rochester, NY: Lawyers' Cooperative, 1962–), 59. On administrative standing and environmental activists, see Susan R. Schrepfer, "Establishing Administrative 'Standing': The Sierra Club and the Forest Service, 1897–1956," in *American Forests: Nature, Culture, and Politics*, ed. Char Miller (Lawrence: University Press of Kansas, 1997), 125–142.

13. *Scenic Hudson Preservation Conference v. Federal Power Commission*, 354 F.2d 608 (2d Cir 1965), *cert denied*, 384 U.S. 941 (1966). See also Arnold W. Reitze Jr., *Air Pollution Control*

Law: Compliance and Enforcement (Environmental Law Institute, Washington, DC, 2001), 599; *Association of Data Processing Services v. Camp,* 397 U.S. 150, 153–154 (1970); quoted in David B. Gregory, *Standing to Sue in Environmental Litigation in the United States of America* (Merges, Switzerland: International Union for Conservation of Nature and Natural Resources, 1972), 12; *United States et al. v. Students Challenging Regulatory Agency Procedures (*SCRAP), 412 U.S. 669 (1973). See also *Sierra Club v. Morton,* 405 U.S. 727 (1972); George Hoberg, *Pluralism by Design: Environmental Policy and the American Regulatory State* (New York: Praeger, 1992), 49–50.

14. Lawrence Pederson, "Anyone Can Be Attorney General," *The Missoulian* (December 15, 1972); "Environment: A New Right to Sue Polluters," *Time* (August 24, 1970): 37.

15. Sax quote from *Defending the Environment: A Strategy for Citizen Action* (New York: Alfred A. Knopf, 1971), 57; see also Sax, "The Public Trust Doctrine in Natural Resource Law," *Michigan Law Review* 68 (1970): 471; Sax, "The Unhappy Truth about NEPA," *Oklahoma Law Review* 26 (1973): 239–248; Sax, "Liberating the Public Trust Doctrine from its Historical Shackles," *University of Davis Law Review* 14 (1980): 185–194; Sax, "Do Communities Have Rights?" *University of Pittsburgh Law Review* 45 (1984): 499–511. For environmentalists' use of Sax, see the "Suggested Reading" list in "Community Action for Environmental Health Education," January 10, 1972, reprinted in GASP's "Leadership Seminar for Citizen Action," October 1972, ff "Leadership Seminar," GASP Internal.

16. Quote from Sax, *Defending the Environment,* 125; cf. Louis L. Jaffe, "Standing to Sue in Conservation Suits," in *Law and the Environment,* ed. Malcolm F. Baldwin, and J. K. Page, Jr. (New York: Walter, 1970); Hoberg, *Pluralism by Design,* 24–25. On the rise of the PIRGs, see Christopher J. Bosso, *Environment, Inc.: From Grassroots to Beltway* (Lawrence: University Press of Kansas, 2005), 64.

17. "The Port Huron Statement," in Alexander Bloom and Wini Breines, eds., *Takin' It to the Streets: A Sixties Reader* (New York: Oxford University Press, 2003), 55; Meta Mendel-Reyes, *Reclaiming Democracy: The Sixties in Politics and Memory* (New York: Routledge, 1995), 32; Robert Fisher, *Let the People Decide: Neighborhood Organizing in America* (New York: Twayne, 1994), 107, 139.

18. "Radical Saul Alinsky: Prophet of Power to the People," *Time* (March 2, 1970); Gordon Whitman, "Beyond Advocacy: The History and Vision of the PICO Network," *Social Policy* 37 (Winter 2006): 50–59.

19. E. L. Stockton, "Pittsburgh: From Smoky City to Smog-less Skies," *UNESCO Courier* 24 (July 1971): 14–17; *A Breath of Fresh Air,* see n. 8; Madoff Earth Day speech, April 22, 1970, AIS Pelkofer; cf. Mary L. Dudziak, *Cold War Civil Rights: Race and the Image of American Democracy* (Princeton: Princeton University Press, 2000), 11–15.

20. "Cinderella City: How Community Action Transformed Pittsburgh's Smoke-Stained Identity," National Association of Manufacturers, *Current Issue Series* 12 (December 1962): 7, ff "Misc. Reports, External, 1962–1977," AIS ACHD BAPC. See commentary on this public relations campaign in Sherie R. Mershon, "Corporate Social Responsibility and Urban Revitalization: An Organizational History of the Allegheny Conference on Community Development, 1943–1990" (Ph.D. dissertation, Carnegie Mellon University, 2000), 739; Nahum Z. Medalia, "Citizen Participation and Environmental Health Action: The Case of Air Pollution Control," *American Journal of Public Health* (August 1969): 1385–1391.

21. "Horizons on Display: Part of the Continuing American Revolution, A Catalogue of Achievement," US GPO 023-000-0031-1, CLP GASP; Marilyn S. Cossetti to Patricia Pelkofer, December 8, 1975; HUD News press release, May 19, 1976; Bicentenntial flag, AIS Pelkofer.

22. *Schoolhouse Rock! 30th Anniversary Collection,* DVD, Walt Disney Studios, 2008.

23. "A Different Cup of Tea," *Time* (December 31, 1973); Christopher Capozzola, "It Makes You Want to Believe: Celebrating the Bicentennial in an Age of Limits," in *America in the Seventies*, ed. David Farber and Beth Bailey (Lawrence: University Press of Kansas, 2004), 32, 36.

24. See Fisher, *Let the People Decide*, xviii; Theda Skocpol and Morris P. Fiorina, "Making Sense of the Civic Engagment Debate," in *Civic Engagement in American Democracy*, ed. Skocpol and Fiorina (Washington, DC: Brookings Institution Press, 1999), 15; Schudson, *The Good Citizen*, 261.

25. "Stenographic Report . . . Pittsburgh, Pennsylvania Monday January 17, 1972, at 9 a.m.," 94, ff "Implementation Plan Hearing, Jan. 17, 1972—Pittsburgh," PSA RG 43.36.

26. Richard Munton, "Deliberative Democracy and Environmental Decision-Making," in *Negotiating Environmental Change: New Perspectives from Social Science*, ed. Frans Berkhout, Melissa Leach, and Ian Scoones (Cheltenham, UK: Edward Elgar, 2003), 109.

27. Schudson, *The Good Citizen*, 264; Hoberg, *Pluralism by Design*, 19–69, especially 31 and 49–50; Frank Uekoetter, *The Age of Smoke: Environmental Policy in Germany and the United States, 1880–1970* (Pittsburgh: University of Pittsburgh Press, 2009), 209.

28. Edward C. Banfield and James Q. Wilson, *City Politics* (Cambridge, MA: Harvard University Press, 1967), 259.

29. Fisher, *Let the People Decide*, 104–118.

30. Robert Gottlieb, *Forcing the Spring: The Transformation of the American Environmental Movement* (Washington, DC: Island Press, 1993), 94; cf. Theda Skocpol, *Diminished Democracy: From Membership to Management in American Civil Life* (Norman: University of Oklahoma Press, 2003), 136–138; Schudson, *The Good Citizen*, 255; Aldon D. Morris, *The Origins of the Civil Rights Movement: Black Communities Organizing for Change* (New York: Free Press, 1984), 288.

31. Sallie A. Marston, "Citizen Action Programs and Participatory Politics in Tucson," in *Public Policy for Democracy*, ed. Helen Ingram and Steven Rathgeb Smith (Washington, DC: Brookings Institution, 1993), 119–135. The EOA is found at Title II of Public Law 88–452, August 20, 1964, the "Economic Opportunity Act of 1964," 78 Stat. 508. Compare this with the development of the "Model Cities" programs: Roy Lubove, *Twentieth Century Pittsburgh*, vol. 2, *The Post-Steel Era* (Pittsburgh: University of Pittsburgh Press, 1996), 87. Quote from Fisher, *Let the People Decide*, 130, and see also 98–167; Schudson, *The Good Citizen*, 257. For VISTA, see Harry C. Boyte, Heather Booth, and Steve Max, *Citizen Action and the New American Populism* (Philadelphia: Temple University Press, 1986), 27–68; Harry C. Boyte, *Everyday Politics: Reconnecting Citizens and Public Life* (Philadelphia: University of Pennsylvania Press, 2004), 23.

32. For the AQA, see § 103 (e) of Public Law 90–148, November 21, 1967, the "Air Quality Act of 1967," 81 Stat. 485; and requirement for public hearings on air quality criteria in § 108 (c) (1). On the Corps of Engineers, see Daniel A. Mazmanian and Jeanne Nienbar Clarke, *Can Organizations Change? Environmental Protection, Citizen Participation and the Corps of Engineers* (Washington, D.C.: Brookings Institute, 1979), especially 132–157; Michael E. Kraft, "U.S. Environmental Policy and Politics: From the 1960s to the 1990s," *Journal of Policy History* 12 (2000): 26; for NEPA and the EIS see § 102(c) of Public Law 91–190, January 1, 1970, the "National Environmental Policy Act of 1969," 83 Stat. 852; for more on the role of the public in the EIS process, see Jack Lewis, "EPA and the Public: A Long Relationship," *EPA* Journal 11 (December 1985): 3–6; § 110 (a)(1) of Public Law 91-604, December 31, 1970, the "Clean Air Act Amendments of 1970," 84 Stat. 1676; Title I, §101 (e) of 92 Public Law 500; "Federal Water Pollution Control Act Amendments of 1972," October 18, 1972, 86 Stat. 816; see comments on this development in GASP's quarterly newsletter, the *Hotline* (October 1973), 8. Issues of the

Hotline are available from CLP GASP, AIS Madoff, and AIS Pelkofer, and have now been made available online at http://www.gasp-pgh.org/hotline/hotline-archive/ .

33. Hoberg, *Pluralism by Design*, 33.

34. Shawn Bernstein, "The Rise of Air Pollution Control as a National Political Issue" (Ph.D. dissertation, Columbia University, 1986), 210.

35. Scott H. Dewey, *Don't Breathe the Air: Air Pollution and U.S. Environmental Politics, 1945–1970* (College Station, TX: Texas A&M University Press, 2000), 73–76, 97, 106–107 and 199–200; Chip Jacobs and William Kelly, *Smogtown: The Lung Burning History of Pollution in Los Angeles* (Woodstock, NY: Overlook Press, 2008), 194–197, 215–221; cf. Gottlieb, *Forcing the Spring*, 109.

36. § 110 (a)(1) of Public Law 91-604, December 31, 1970, the "Clean Air Act Amendments of 1970," 84 Stat. 1676.

37. Lester B. Lave and Gilbert S. Omenn, *Clearing the Air: Reforming the Clean Air Act* (Washington, DC: Brookings Institution, 1981), 7.

38. Community Support Bulletin, National Air Pollution Control Administration, Office of Education and Information, "How Important Is the Public in Public Hearing?" January 15, 1970, ff "Research Papers and Reports," AIS Broughton.

39. Loretta A. [illegible], Northside Environmental Action Committee, to William D. Ruckelshaus, Indianapolis, IN, January 2, 1972; Stanley H. Vegors, Jr., to John T. Middleton, Pocatello, ID, May 2, 1972; Lowella H. "Bunny" Butler, Group Against Smog and Pollution, to Ruckelshaus, El Paso, TX, March 24, 1972; Ralph Madison, Clean Air Committee of the Louisville Audubon Society, to U.S. EPA, Louisville, KY, February 17, 1972; William Darrah, to Ruckelshaus, Hawaii, February 3, 1972; William D. Tomlinson, to Ruckelshaus, Missoula, Montana, February 26, 1972. All of these letters from ff "State Air Standards," Office of Air Programs Administrative Subject Files, 1966–1972, NARA RG 412.

40. Citizen's Advisory Committee on Environmental Quality, "Community Action for Environmental Quality" (April 1970), quotes from 26 and 31; Hoberg notes that Health, Education, and Welfare and EPA encouraged citizen involvement in this way for only three years before opposition shuttered the program. Hoberg, *Pluralism by Design*, 61.

41. "Citizen Action Can Get Results" (August 1972), U.S. GPO 1972 0-469-955, 8; cf. "The Citizen's Role in Environmental Decision Making" (November 1972), U.S. GPO 1972 0-478-748; GASP commentary on latter in GASP *Hotline* (March–April 1973), 7.

42. John Middleton, deputy assistant administrator, to EPA Office of the Administrator, June 7, 1971, ff "OC Citizen's Committee 1971," Office of Air Programs Administrative Subject Files, 1966–1972, NARA RG 412.

43. Ibid; see also EPA Office of the Administrator to John Middleton, deputy assistant administrator, June 2, 1971, Office of Air Programs Administrative Subject Files, 1966–1972, NARA RG 412.

44. The PSAPCA's position in the state of Washington was much like the Allegheny County Health Department's role as the sole local air pollution control agency in a state otherwise regulated by the Pennsylvania Department of Environmental Resources.

45. James Pearson, "Indirect Source Review: Problems for the Air Pollution Control Agency," *Journal of the Air Pollution Control Association* 28 (1978): 367–370.

46. James Pearson, *Reducing the Averages: The Founding and Development of the Puget Sound Air Pollution Control Agency* (San Jose, CA: Writer's Club Press, 1999), 146.

47. Ibid., 145.

48. See Kenneth Edward Melichar, "The Making of the 1967 Montana Clean Air Act: A Struggle over the Ownership of Definitions of Air Pollution," *Sociological Perspectives* 30

(1987): 49–70; Melichar, "The Making of the 1967 Montana Clean-Air-Act: A Sociological Study of Environmental Legislation" (Ph.D. dissertation, New York University, 1985).

49. Arthur Hutchinson, "Implementation Plan Status in Doubt after Disagreement," *Sunday Missoulian* (February 6, 1972): 24.

50. The first cochair of Montana GASP was Nancy Fritz; the secretary was Margaret E. Schaefer. See Gary Langley, "GASPer Chronicles Pollution Fight," *Missoulian* (December 15, 1971); Margaret E. Schaefer to Edmund S. Muskie, Missoula, Montana, February 12, 1972, ff "State Air Standards," NARA RG 412. Despite the similarities between GASP in Pittsburgh and Montana's GASP, there is no specific evidence of direct contact between the two organizations before the creation of Montana GASP in 1971. In 1972, however, one Missoula resident attended a Pittsburgh GASP leadership seminar. "GASP Leadership Seminar for Citizen Action: Participant Profiles," GASP pamphlet, October 1972, ff "Leadership Seminar," GASP Internal.

51. Lawrence Pederson, "Anyone Can Be Attorney General," *Missoulian* (December 15, 1972).

52. Mr. and Mrs. John D. Grove to Ruckelshaus, January 10, 1972, ff "State Air Standards," NARA RG 412.

53. Albert Smith testimony before the Environmental Quality Board, January 1972, PSA RG 43.16.

54. National Environmental Policy Act of 1969, 42 U.S.C. 4321–4361; quotation at Title 1, § 102 (c).

55. Serge Taylor, *Making Bureaucracies Think: The Environmental Impact Statement Strategy of Administrative Reform* (Stanford, CA: Stanford University Press, 1984), 5, 7, and 295; Samuel P. Hays, *Beauty, Health and Permanence: Environmental Politics in the United States, 1955–1985* (Cambridge: Cambridge University Press, 1987), 474; cf. Munton, "Deliberative Democracy," 109.

56. Hoberg, *Pluralism by Design*, 199–198.

57. Andrew Hurley, *Environmental Inequalities: Class, Race and Industrial Pollution in Gary, Indiana, 1945–1980* (Chapel Hill: University of North Carolina Press, 1995), 56.

58. For more on the importance of the League of Women Voters as a "seed" organization for environmental activism, see Hurley, *Environmental Inequalities*, 57–60; Terence Kehoe, *Cleaning Up the Great Lakes: From Cooperation to Confrontation* (Dekalb: Northern Illinois University, 1997), 52–53; Hays, *Beauty, Health and Permanence*, 460–461.

59. Beth Irons to Sheldon W. Samuels, March 3, 1971, ff "OC Citizen's Committee 1971 (1)," NARA RG 412; "E.P. Group to Discuss," El Paso (TX) *Herald-Post*, September 29, 1971; Butler to Ruckelshaus, March 24, 1972.

60. Harriet Marble to Ruckelshaus, February 10, 1972, NARA RG 412.

61. Hurley, *Environmental Inequalities*, 57–60.

62. Jacobs and Kelly, *Smogtown*, 240–241.

63. Mrs. Eugene M. Landis to EQB, December 28, 1971, ff "Correspondence RE: Air Hearings, December 1, 2, 3, 4 1971," PSA RG 43.37.

64. Hoberg, *Pluralism by Design*, 29; Matthew A. Crenson, *The Un-Politics of Air Pollution: A Study of Non-decisionmaking in the Cities* (Baltimore: Johns Hopkins University Press, 1971).

65. "Do's and Don't's [*sic*] for testimony at public hearings," ff "Training Workshops." GASP's specific recommendations were borrowed from the "Delaware Citizens for Clean Air" newsletter; "Procurement Plan for Community Air Implementation Workshops," ff "Public Meetings"; "Citizen's Workshop on Implementation," ff Testimony Workshops, all from GASP Internal.

66. "Pollution Hearings," *PF* 1/21/72.

67. "Tentative Schedule for Testimony . . . Harrisburg, December 2, 1971," ff "State IP," GASP Internal; December 16, 1971, 3; "Environmental Quality Board, Minutes 1971–1986," ff "Minutes," PSA RG 43.36.

68. "Wife Wars against Pollution," *PP* 12/2/71; "After 25 Years," *PP* 11/23/71; "Pollution Hearings," *PF* 1/21/72; "60 Request Time," *PP* 11/28/71; "State Weighs What Powers," *PP* 12/2/71; "Air Code Hearing," *PP* 12/2/71; "Gasp! Clean Air Talks Here," *PP* 12/3/71; "Airing the Problem," *PPG* 11/30/71; "Witness Lineup Extends Session," *PPG* 12/1/71; "GASP Issues Alert," *PPG* 12/3/71.

69. "Verbatim Report of Hearing Held in Coughlin High School Building, . . . September 26, 1961"; "Verbatim Report of Hearing Held in Clarion, . . . September 21, 1961"; "Verbatim Report of Hearing Held in Washington County . . . September 20, 1961," ff "Transcripts of Hearings on Proposed Regulations, 1961, Regulation I," PSA 11.43.

70. "Air Pollution Hearing: Pittsburgh Pennsylvania, Tuesday September 9, 1969," quotes from pages 51, 79, 150, 153, 246, PSA RG 11.43.

71. "State Chamber's Air Quality Stand," *PP* 12/3/71; "Industry's Interest," *PP* 11/7/71.

72. "State Ends Clean Air Probe," *PP* 12/5/71.

73. "Statement presented at Environmental Quality Board Meeting, Harrisburg Pennsylvania, Thursday, January 27, 1972, by Michelle Madoff," 1, ff "State IP," GASP Internal. Emphasis in original.

74. "Implementation Plan Hearing, January 17, 1972, Pittsburgh," 207, 216, and 115, PSA RG 43.37.

75. Lubove, *Twentieth Century Pittsburgh*, 59; Skocpol and Fiorina, "Making Sense of the Civic Engagement Debate," 2; Richard N. L. Andrews, "Class Politics or Democratic Reform: Environmentalism and American Political Institutions," *Natural Resources Journal* 20 (1980): 222; cf. Andrews, *Managing the Environment, Managing Ourselves: A History of American Environmental Policy* (New Haven: Yale University Press, 1999), 209.

Chapter 2: The Smoky City

1. Edward K. Muller and Joel A. Tarr, "The Interaction of Natural and Built Environments in the Pittsburgh Landscape," in *Devastation and Renewal: An Environmental History of Pittsburgh and Its Region*, ed. Joel A. Tarr (Pittsburgh: University of Pittsburgh Press, 2003), 11–14; R. E. Harper, *The Transformation of Western Pennsylvania, 1770–1800* (Pittsburgh: University of Pittsburgh Press, 1991), 90; Walter O'Meara, *Guns at the Forks* (Englewood Cliffs, NJ: Prentice-Hall, 1965), 252.

2. George Rogers Taylor, *The Transportation Revolution, 1815–1860* (New York: Rinehart, 1951), 387.

3. Frederick Moore Binder, *Coal Age Empire: Pennsylvania Coal and Its Utilization to 1860* (Harrisburg: Pennsylvania Historical and Museum Commission, 1974), 42; Roy Lubove, *Twentieth Century Pittsburgh*, vol. 2, *The Post-Steel Era* (Pittsburgh: University of Pittsburgh Press, 1996), 30.

4. Quote from James Parton, "Pittsburgh," *Atlantic Monthly* 21 (1868): 17. The cramped conditions in Pittsburgh created logistical difficulties for industrial development; see Kenneth Warren, *The American Steel Industry, 1850–1970: A Geographical Interpretation* (Oxford: Clarendon Press, 1973), 134–138; Warren, *Big Steel: The First Century of the United States Steel Corporation, 1901–2001* (Pittsburgh: University of Pittsburgh Press, 2001), 266–267; David Stradling, *Smokestacks and Progressives: Environmentalists, Engineers and Air Quality in America, 1881–1951* (Baltimore: Johns Hopkins University Press, 1999), 23; Robert Dale Grinder, "The

Battle for Clean Air: The Smoke Problem in Post–Civil War America," in *Pollution and Reform in American Cities, 1870–1930*, ed. Martin V. Melosi (Austin: University of Texas Press, 1980), 84. The Londoners were not known as "inversions" until the twentieth century. See Sumner B. Ely, "General Atmospheric Pollution: Some Works in Pittsburgh Air Pollution Problems," *American Journal of Public Health* 38 (July 1948): 960–961.

5. Quote from *Thomas's Travels through the Western Country in 1816*, reprinted in *Pittsburgh in 1816: Compiled By the Carnegie Library of Pittsburgh on the One Hundredth Anniversary of the Granting of the City Charter* (Pittsburgh: Carnegie Library, 1916), 13. Cf. T. J. Chapman, *Old Pittsburgh Days* (Pittsburgh: J.R. Weldin, 1900), 191. For more on early nineteenth century Pittsburgh coal smoke, see Binder, *Coal Age Empire*, 21. Binder cites an 1800 traveler's comments from H. N. Eavenson, "The Early History of the Pittsburgh Coal Bed," *Western Pennsylvania Historical Magazine* 22 (1939): 167–174.

6. Thomas J. Misa, *A Nation of Steel: The Making of Modern America, 1865–1925* (Johns Hopkins University Press, 1995), 177; quote from "Pittsburg," *Harper's Weekly* (February 18, 1871): 147.

7. For the nation's "Coal Age," see Stradling, *Smokestacks and Progressives*, 7, 12; David J. Cuff and William J. Young, eds., *The Atlas of Pennsylvania* (Philadelphia: Temple University Press, 1989), 10 and map, 14–15. For a further geographical description of the Pittsburgh Seam, see Kenneth Warren, *Waste, Wealth, and Alienation: Growth and Decline in the Connelsville Coke Industry* (Pittsburgh: University of Pittsburgh Press, 2001), 13. A layman's explanation of the chemistry of coal smoke is available from Barbara Freese, *Coal: A Human History* (Cambridge, MA: Perseus, 2003), 166–177.

8. On the interaction of geography and meteorology, see Michael J. McGraw and Charles S. Holt, National Air Pollution Control Administration, "Pittsburgh Metropolitan Area Air Pollutant Emission Inventory" (November 1968), 10, ff "Pittsburgh," Administrative Subject Files, 1966–1972, NARA RG 412; Stradling, *Smokestacks and Progressives*, 23–24; Donald B. Dodd, ed., *Historical Statistics of the States of the United States: Two Centuries of the Census, 1790–1990* (Westport, CT: Greenwood Press, 1993), 455–456.

9. "The City of Pittsburg," *Harper's Weekly* (February 27, 1892): 202.

10. Stradling, *Smokestacks and Progressives*, 62; Noga Morag-Levine, *Chasing the Wind: Regulating Air Pollution in the Common Law State* (Princeton: Princeton University Press, 2003), 86–102.

11. Jacqueline Karnell Corn, *Environment and Health in Nineteenth Century America: Two Case Studies* (New York: Peter Lang, 1989), 207; "The Allegheny County Health Department: Present Capacity and Future Potential," in *Report of the Evaluation Assessment Team of the Graduate School of Public Health, University of Pittsburgh* (Pittsburgh: University of Pittsburgh Press, 1992), appendix A, 2–3. St. Louis's 1867 ordinance preceded Pittsburgh's 1869 law. Joel A. Tarr and Carl Zimring, "The Struggle for Smoke Control in St. Louis: Achievement and Emulation," in *Common Fields: An Environmental History of St. Louis*, ed. Andrew Hurley (St. Louis: Missouri Historical Society Press, 1997), 203; Morag-Levine, *Chasing the Wind*, 103; Stradling, *Smokestacks and Progressives*, 38–39; Joel A. Tarr, *Search for the Ultimate Sink: Urban Pollution in Historical Perspective* (Akron, OH: University of Akron Press, 1996), 231.

12. Robert H. Wiebe, *The Search for Order, 1877–1920* (New York: Hill and Wang, 1986); Samuel P. Hays, *Conservation and the Gospel of Efficiency* (Cambridge: Cambridge University Press, 1958).

13. Quote from Carolyn Merchant, "Women of the Progressive Conservation Movement, 1900–1916," *Environmental Review* (Spring 1984): 57; Tarr and Zimring, "The Struggle for Smoke Control," 202; Grinder, "Battle," 88; Harold L. Platt, "Invisible Gases: Smoke, Gender,

and the Redefinition of Environmental Policy in Chicago, 1900–1920," *Planning Perspectives* (1995): 67; Maureen A. Flanagan, "Gender and Urban Political Reform: The City Club and the Women's City Club of Chicago in the Progressive Era," *American Historical Review* (1990): 1048; Suellen M. Hoy, "'Municipal Housekeeping': The Role of Women in Improving Urban Sanitation Practices," in Melosi, *Pollution and Reform in American Cities*, 194.

14. Loretta S. Lobes, "Hearts All Aflame: Women and the Development of New Forms of Social Service Organization, 1870–1930" (Ph.D. dissertation, Carnegie Mellon University, 1996), 109; Marie Dermitt, *Fifty Years of Civic History* (Pittsburgh: Civic Club of Allegheny County, 1945), 5; John F. Bauman and Margaret Spratt, "Civic Leaders and Environmental Reform: The Pittsburgh Survey and Urban Planning," in *Pittsburgh Surveyed: Social Science and Social Reform in the Early Twentieth Century*, ed. Maurine W. Greenwald and Margo Anderson (Pittsburgh: University of Pittsburgh Press, 1996), 155–157. For more on the impact of natural gas on the air quality of Pittsburgh, see Angela Gugliotta, "How, When, and for Whom Was Smoke a Problem in Pittsburgh?" in Tarr, *Devastation and Renewal*, 114–116; Joel A. Tarr and Bill C. Lamperes, "Changing Fuel Use Behavior and Energy Transitions: The Pittsburgh Smoke Control Movement, 1940–1950," *Journal of Social History* 14 (1981): 561–588 562, 576.

15. David Stradling, "Civilized Air: Coal, Smoke and Environmentalism in America, 1880–1920," (Ph.D. dissertation, University of Wisconsin–Madison, 1996), 90–91.

16. Quotes from Lobes, "Hearts All Aflame," 110–113.

17. Nora Faires, "Immigrants and Industry: Peopling the 'Iron City'" in *City at the Point: Essays on the Social History of Pittsburgh*, ed. Samuel P. Hays (Pittsburgh: University of Pittsburgh Press, 1989), 3–32.

18. Joel A. Tarr, "Infrastructure and City-Building in the Nineteenth and Twentieth Centuries," in Hays, *City at the Point*, 241–242; Tarr, "The Pittsburgh Survey as an Environmental Statement," in Greenwald and Anderson, *Pittsburgh Surveyed*, 179–181; Stradling, *Smokestacks and Progressives*, 98–100; Ralph H. German, "Regulation of Smoke and Air Pollution in Pennsylvania," *University of Pittsburgh Law Review* 10 (1948–1949): 495; Charles O. Jones, *Clean Air: The Policies and Politics of Pollution Control* (Pittsburgh: University of Pittsburgh Press, 1975), 22; Roy Lubove, *Twentieth-Century Pittsburgh*; vol. 1, *Government, Business and Environmental Change* (Pittsburgh: University of Pittsburgh Press, 1995), 46–48.

19. Stradling, *Smokestacks and Progressives*, 76–77.

20. Martin V. Melosi, *Coping with Abundance: Energy and the Environment in Industrial America* (Philadelphia: Temple University Press, 1985), 33; Langdon White, "The Iron and Steel Industry of the Pittsburgh District," *Economic Geography* (1928): 115–139; White, "Geography's Part in the Plant Cost of Iron and Steel Production at Pittsburgh, Chicago, and Birmingham," *Economic Geography* (1929): 327–334; Lubove, *Twentieth-Century Pittsburgh*, vol. 1, 114.

21. Tarr, *Search for the Ultimate Sink*, 563.

22. Ibid., 564, 525.

23. For more on the Ringelmann Chart, see Scott H. Dewey, *Don't Breathe the Air: Air Pollution and U.S. Environmental Politics, 1945–1970* (College Station: Texas A&M University Press, 2000), 118–119; Stradling, *Smokestacks and Progressives*, 72; "Maximilian Ringelmann—Man of Mystery," *Air Repair* (November 1952): 4–6; Frank Uekoetter, *The Age of Smoke: Environmental Policy in Germany and the United States, 1880–1970* (Pittsburgh: University of Pittsburgh Press, 2009), 159, 266. For an insider's assessment, see James Pearson, *Reducing the Averages: The Founding and Development of the Puget Sound Air Pollution Control Agency* (San Jose, CA: Writer's Club Press, 1999), 24–25.

24. Quotes from Sherie R. Mershon, "Corporate Social Responsibility and Urban Revitalization: An Organizational History of the Allegheny Conference on Community Development,

1943–1990" (Ph.D. dissertation, Carnegie Mellon University, 2000), 20, 3, 297. On the 1941 law, see Dewey, *Don't Breathe the Air*, 32. The USC, an umbrella group consisting of eighty organizations, was created by the Civic Club and only later merged with the ACCD. Tarr, *The Search for the Ultimate Sink*, 244–245. For more on the USC and the 1949 law, see Mershon and Tarr, "Strategies for Clean Air: The Pittsburgh and Allegheny County Smoke Control Movements, 1940–1960," in Tarr, *Devestation and Renewal*, 145–173.

25. "Smoke Sleuths: Pittsburgh Begins an All-Out War on Its Old Enemies, Smoke and Smog," *Life* (December 1, 1947): 147; William G. Willis, *The Pittsburgh Manual: A Guide to the Government of the City of Pittsburgh* (Pittsburgh: University of Pittsburgh Press, 1950), 140–143; Bruce M. Stave, *The New Deal and the Last Hurrah: Pittsburgh Machine Politics* (Pittsburgh: University of Pittsburgh Press, 1970), 3–23, 183–194; Michael P. Weber, *Don't Call Me Boss: David L. Lawrence, Pittsburgh's Renaissance Mayor* (Pittsburgh: University of Pittsburgh Press, 1988), quotation on 235, see also 214, 228–247; Lubove, *Twentieth-Century Pittsburgh*, vol. 1, 114.

26. Lubove, *Twentieth-Century Pittsburgh*, vol. 1, 121.

27. Advisory committee quote from Jones, *Clean Air*, 46; Edward L. Stockton, "The Pittsburgh Air Pollution Control Story," *American Industrial Hygiene Association Journal* (September–October 1966): 469–474. Cf. Tarr and Zimring, "The Struggle for Smoke Control in St. Louis," 218.

28. Joel A. Tarr, *Search for the Ultimate Sink*, 561; Film *Report from America: Smokey Pittsburgh*, BBC, 1952. In the video collection of Hunt Library, Carnegie Mellon University.

29. Mershon and Tarr, "Strategies for Clean Air," 171. County health departments were created under the Pennsylvania Legislature's "Local Health Administration Law," Act 315 of 1951. See "The Allegheny County Health Department: Present Capacity and Future Potential," appendix A, 2–4.

30. Commonwealth of Pennsylvania, Air Pollution Control Act, 1960, § 2; quotes from Jones, *Clean Air*, 50-51; collected minutes of the Advisory Committee, Department of Health Air Pollution Commission "Meeting Packets, 1961–1971," PSA RG 11.42. This limited public representation was the industrial ideal of the time—see the special issue of *Power* 104 (December 1960).

31. Dewey, *Don't Breathe the Air*, 239–248; for an overview of the Clean Air Acts and Amendments, see Lester B. Lave and Gilbert S. Omenn, *Clearing the Air: Reforming the Clean Air Act* (Washington, DC: The Brookings Institution, 1981).

32. Barbara Ferman, *Challenging the Growth Machine: Neighborhood Politics in Chicago and Pittsburgh* (Lawrence: University Press of Kansas, 1996), 86–87.

33. Richard L. Forstall, *Population of Counties by Decennial Census: 1900 to 1990* (Washington, DC: Population Division, US Bureau of the Census, 1995); *Allegheny County Government: Organization, Facilities and Services* (Pittsburgh: League of Women Voters, 1971), 6–7.

34. Ferman, *Challenging the Growth Machine*, 159 n. 22.

35. See Mershon, "Corporate Social Responsibility"; Gregory J. Crowley, *The Politics of Place: Contentious Urban Redevelopment in Pittsburgh* (Pittsburgh: University of Pittsburgh Press, 2005), 36–49; Lubove, *Twentieth Century Pittsburgh*, vol. 2, chapter 3.

36. Mershon, "Corporate Social Responsibility," 707.

37. Brian Apelt, *The Corporation: A Centennial Biography of the United States Steel Corporation, 1901–2001* (Pittsburgh: Cathedral, 2000), 335.

38. For a national perspective, see Dewey, *Don't Breathe the Air*, 85 and 256 n. 10.

39. Richard N. L. Andrews, *Managing the Environment, Managing Ourselves: A History of American Environmental Policy* (New Haven: Yale University Press, 1999), 210–218; Dewey, *Don't Breathe the Air*, 155–156. For more on Donora—a western Pennsylvania mill town in

which a week-long temperature inversion trapped pollutants from the local zinc mill, sickening many and killing eighteen in 1948—see Devra Davis, *When Smoke Ran like Water: Tales of Environmental Deception and the Battle against Pollution* (New York: Basic, 2003), 1–30.

40. Donald P. Eriksen, "Air Pollution: The Problem in the Legislative, Administrative Responses of the United States, Pennsylvania, and Allegheny County," *University of Pittsburgh Law Review* 30 (1969): 663–662. This article predates Article XVII revisions. The Local Health Administration Law is found at Pa. Stat. Ann., 16 § 12010 (f) (1956).

41. "County OKs 'Toughest' Smoke Bill," *PPG* 6/7/60; "'Toughest' Smoke Law Promised," *PP* 6/21/59; "Toughest Smoke Law Voted," *PP* 5/7/60; not to mention "County Vows 'Toughest' Laws," Pittsburgh *Sun-Telegraph* 6/17/59. The word "tough" showed up again in journalistic description of Article XIII's successor, Article XVII: "Tough Clean-Air Code Unveiled in Pittsburgh," *Washington Post* 11/29/69; Joy Flowers Conti and Janice L. Gambino, "Local Regulation of Air Pollution: The Allegheny County Experience," *Duquesne Law Review* (1973): 617.

42. Mrs. Thomas Horrocks to Dr. John D. Lauer, April 20, 1960, ff 1, AIS ACAPCAC.

43. "Housewife Gets Post," *PP* 9/14/60; see also "County Mother Cited," *PP* 6/23/69.

44. Mershon, "Corporate Social Responsibility," 737; "Steeler's [*sic*]-Saints Football Game, October 20, 1968," ACHD BAPC.

45. Minutes of Meeting, Allegheny County Air Pollution Control Advisory Committee, November 3, 1965, ff 1, AIS ACAPCAC; cf. "County Mother Cited," *PP* 6/23/69.

46. Among many, see "'Cleaner Air' Accent Starts," *MDN* 10/ 21/68.

47. Nahum Z. Medalia, "Citizen Participation and Environmental Health Action: The Case of Air Pollution Control," *American Journal of Public Health* (August 1969): 1386.

48. "Cinderella City: How Community Action Transformed Pittsburgh's Smoke Stained Identity," National Association of Manufacturers, *Current Issues* series 12, December 1962, in ff "17," AIS ACHD BAPC.

49. "The Pittsburgh Air Pollution Control Story," presented at the 1966 American Industrial Hygiene Conference, May 18, 1966, CLP GASP.

50. Shawn Bernstein, "The Rise of Air Pollution Control as a National Political Issue" (Ph.D. dissertation, Columbia University, 1986), 232.

51. See Samuel P. Hays, *Beauty, Health and Permanence: Environmental Politics in the United States, 1955–1985* (Cambridge: Cambridge University Press, 1987), 73–74, for a discussion of the evolution of federal and state interdependence in the 1963, 1967, and 1970 Clean Air Acts; ACHD press release, August 25, 1969, in ff "Allegheny County Bureau of Air Pollution Control—Press Releases," AIS Broughton. Quote from the Air Quality Act of 1967, P.L. 90-148, November 21, 1967, 81 Stat. 485; Jones, *Clean Air*, 156–157.

52. NAPCA, "Community Support Bulletin: "How Important Is the Public in Public Hearings?" (January 15, 1970). AIS Broughton.

53. "The 'Air War,'" *WSJ* 10/21/69.

54. "County Racing Date," *PP* 9/28/69.

55. "Keeping Pace with Environmental Issues: Public Turns Out for Air Pollution Hearings," *Water Land and Life: A Publication of Western Pennsylvania Conservancy* (December 1969): 6, in ff "Historic GASP Materials (Misc)," AIS Pelkofer.

56. "Air Pollution Hearing: Pittsburgh Pennsylvania, Tuesday September 9, 1969," 68, 184–185, 134, 136, ff 1, PSA RG 11.43; "Mrs. J. Lewis Scott" is most likely Ruth Joan Scott, personal friend of Rachel Carson, graduate of the Carnegie Institute of Technology, and environmental educator with the Carnegie Museum of Natural History; the transcript appears to have muddled Scott's claim to represent one or the other of these institutions. "Ruth Jury Scott," *PPG* 6/19/03.

57. Madoff to "Fellow Breathers," October 1969, AIS Madoff.

58. See the discussion of the County Air Pollution Advisory Committee, above, and *PP* 9/28/69.

59. "The 'Air War,'" *WSJ* 10/21/69.

60. Note that while Articles IX (1941), XII (1958), and XIII (1960) preceded Articles XVII (1970), XVIII (1972), and XXI (1994), the articles referring to air pollution control in the ACHD Rules and Regulations are not sequential; the article numbers come from the order in which the articles are presented in subsequent revisions of the Rules and Regulations. Therefore, Articles XIV, XV, XVI, XIX, and XX don't refer to air pollution matters. County of Allegheny, Pennsylvania, Ordinance No. 16782, and ACHD Rules and Regulations, Article XXI Air Pollution Control, effective 1994, amended 1995 and 1998. Quote from Allegheny County Rules and Regulations XIII § 1303 part 5.

61. Allegheny County Rules and Regulations IX § 903 part 5, XIII § 1303 part 5, and XVII § 1700, ff 13 "Rules and Regulations, Allegheny County Health Department, 1949–1978," AIS ACHD BAPC. Article XVIII (1972) maintained much of the wording of the previous Rules and Regulations.

62. Quotes from Ferman, *Challenging the Growth Machine*, 38–39, 86–87; Lubove, *Twentieth-Century Pittsburgh*, vol. 2, 58–59.

63. Untitled GASP pamphlet, n.d. (ca. 1970), 4, CLP GASP.

64. For published accounts of GASP, see Jones, *Clean Air*; Robert Gottlieb, *Forcing the Spring: The Transformation of the American Environmental Movement* (Washington, DC: Island Press, 1993), 126–127; John O. Frohliger, "GASP: An Ecological David," *Communiqué: Journal of the University of Pittsburgh Hospitals* (1972): 12–16; Lynton K. Caldwell, Lynton R. Hayes, and Isabel M. MacWhirter, *Citizens and the Environment: Case Studies in Popular Action* (Bloomington: Indiana University Press, 1976), 130–133. The most recent information about GASP is available from the group's website, http://www.gasp-pgh.org/.

65. Madoff credits University of Pittsburgh professor and GASP member Mort Corn with the original idea of an independent board. Telephone interview with Michelle Madoff, February 14, 2002. Tape and transcript in author's possession.

66. For descriptions of the impact of Article XVII and XVIII, see Conti and Gambino, "Local Regulation of Air Pollution," 612–656; Robert S. Bailey, "Air Pollution Control for Allegheny County: Will It Be Smothered by Appellate Procedure?" *Duquesne Law Review* (1970): 395–406; Kenneth L. Hirsch and Steven Abramowitz, "Clearing the Air: Some Legal Aspects of Interstate Air Pollution," *Duquesne Law Review* (Fall 1979): 53–102; and Samuel P. Hays, "Clean Air: From the 1970 Act to the 1977 Amendments," *Duquesne Law Review* (Fall 1978): 33–66. Quote from "Where the Action Is: Variance Appeals," GASP *Hotline* (September 1970), 1, CLP GASP.

67. Telephone interview with author, February 14, 2002.

68. "No Rate Hike," *PP* 1/23/70; "Power Firm Chided," *PPG* 1/23/70.

69. Broughton to David Olds, November 3, 1970, ff 3, AIS Broughton. The utility also petitioned (unsuccessfully) for a continuance: Broughton to Olds July 2, 1970.

70. Olds to Broughton, July 6, 1970, ff 3, AIS Broughton.

71. Allegheny County Health Department Rules and Regulations, Article XVII, January 1, 1970, § 1703.1 F3 and F6.

72. "Duquesne Light's Lawyer Protests," *PP* 8/7/70, also cited in Jones, *Clean Air*, 217–218. It is worthwhile to note that while GASP and Duquesne Light clashed early and often, by 1973 GASP was a vocal supporter of the utility company's innovative attempts to limit sulfur oxide from coal-fired power plants. See "Public Hearing and Conference on Status of Compliance

with Sulfur Oxide Emission Regulations by Power Plants October 18–November 2, 1973," vol. 7, 2978, and vol. 6, 1536 and 1580, NARA RG 412.

73. Gorr to Broughton, July 8 1970, ff 3, AIS Broughton.

74. "Public Wins Role," *PP* 7/20/70.

75. "Duquesne Light's Lawyer Protests," *PP* 8/7/70.

76. "Report to the Air Pollution Control Advisory Committee by Ronald J. Chleboski," June 11, 1970 meeting minutes, ff 6, AIS ACAPCAC.

77. "*In the Matter of U.S. Steel Corporation v. Commonwealth of Pennsylvania Department of Environmental Resources*," docket no. 72-397-D, *Pennsylvania Environmental Hearing Board Adjudications 1975* (Harrisburg: Commonwealth of Pennsylvania, 1979).

78. Editorial, *PP* 7/27/70. Reprinted as "Reply to a Complaint," *PP* 8/7/70.

79. "Apologize to Duquesne Light?" *PP* 8/13/70; GASP *Hotline* (1971): 1, AIS Pelkofer.

80. Esther Kitzes, "GASP Plans Ahead," *PF* 10/30/70.

81. "Foes Fired Up," *PP* 11/1/70; "Pitt Profs Rap Duquesne Light," *PP* 11/2/70; "Lung Peril," *PPG* 11/3/70; "Conservation Group Rips," *PP* 11/23/70.

82. Interview with Charles Raymond Stowell Jr., June 21, 2001. Tape and transcript in author's possession.

83. Allegheny County Regulations, Article XVII, January 1, 1970, § 1703.1F10; *Pennsylvania State Annual*, title 71, § 1710:41; Robert S. Bailey, "Air Pollution Control for Allegheny County—Will It Be Smothered by Appellate Procedure?" *Duquesne Law Review* (1970): 402.

84. For extensive discussions of citizen suits from the legal perspective, see David P. Gionfriddo, "Sealing Pandora's Box: Judicial Doctrines Restricting Public Trust Citizen Environmental Suits," *Boston College Environmental Affairs Law Review* (Spring 1986): 439; Sam Kalen, "Standing on Its Last Legs: *Bennett V. Spear* and the Past and Future of Standing in Environmental Cases," *Journal of Land Use and Environmental Law* (Fall 1997): 1; Carole L. Gallagher, "The Movement to Create an Environmental Bill of Rights: From Earth Day, 1970 to the Present," *Fordham Environmental Law Journal* (Fall 1997): 107; Gene R. Nichol, "Citizen Suits and the Future of Standing in the 21st Century: From Lujan to Laidlaw and Beyond: The Impossibility of Lujan's Project," *Duke Environmental Law and Policy Forum* (Spring 2001): 193.

85. Pat Newman, "Enforcement, Pittsburgh Style, or, Who Took the Teeth out of Article XVIII?" GASP *Hotline "Citizen's Commentary"* (March 1973): 3, ff "Hotlines and Other GASP Publications," GASP Internal; "An Act Amending the 'Air Pollution Control Act,'" Oct. 26, 1972, P.L. 989, No. 245.

86. "Pittsburgh Cracks Down on Polluters," *Business Week* (December 27, 1969): 66; "Will Anti-Pollution Laws Stick?" *WSJ* 6/2/70.

Chapter 3: "I Belong Here!"

1. *I Belong Here!* GASP (1972), 16 mm film, CLP GASP.

2. Press release, "I Belong Here," January 1. 1975, ff 10, GASP Internal.

3. "Between Gasps of Fresh Air," *Christian Science Monitor* (February 20, 1970): 2; "Pittsburgh Cracks Down on Polluters," *Business Week* (December 27, 1969): 66–67; "The 'Air War': 'Breather's Lobby' Wins," *WSJ* 10/21/69.

4. "'Newsroom' Celebrates Its First Year," *PF* 12/70; John O. Frohliger, "GASP: An Ecological David," *Communiqué: Journal of the University of Pittsburgh Hospitals* (1972): 12–16; ff "Awards to GASP, 1970s," GASP Internal; "GASP at 10," *PPG* 12/4/79.

5. "GASP Getting Two Awards," *PP* 10/21/71; "Named by Shapp," *PP* 4/29/71; press release, "I Belong Here."

6. Roy McHugh, "Still the Same Old," *PP* 07/01/72; John Hoerr, "U.S.S. Clairton Works," *PF* 10/29/71; "GASP Plans Ahead," *PF* 10/30/70; Bloom telephone interview with author, September 4, 2007; Madoff quoted in Charles O. Jones, *Clean Air: The Policies and Politics of Pollution Control* (Pittsburgh: University of Pittsburgh Press, 1975), 150. Other journalistic portraits of Madoff include "GASP Expects Action," *PP* 11/3/69; Helen Frank Collins, "Between Gasps of Fresh Air . . ." *Christian Science Monitor* (February 20, 1970): 2.

7. Jones, *Clean Air*, 151.

8. Collins, "Between Gasps of Fresh Air," 2; speech delivered on unknown date, hand notes on typescript, "1 year history," AIS Madoff; ellipsis in original.

9. Madoff interview with author, Surprise, Arizona, October 15–16, 2007; letters to the editor, *PPG* 2/26/75.

10. Bloom quote from telephone interview, spring 2009; for a definition of neighborhood as used here, see Robert Fisher, *Let the People Decide: Neighborhood Organizing in America* (New York: Twayne, 1994), xix.

11. GASP Board of Directors' meeting, February 23, 1970, ff "GASP—Records—early material 1969–1970," HSWP GASP. Thirty-nine separate zip codes, corresponding to Pittsburgh's major neighborhoods, are listed in the United States Post Office Department Postal Zip Code Directory for Pittsburgh, published in April 1969, GPO # 963-111.

12. Bloom quote from telephone interview, spring 2009; while it is now known as Carnegie Mellon University, the school punctuated its name "Carnegie-Mellon" until the late 1990s. The name was changed from Carnegie Institute of Technology in 1967, though newspaper accounts and correspondence from the early 70s often continued to refer to the school by that name, or as Carnegie Tech or the Carnegie Institute. See Edwin Fenton, *Carnegie Mellon 1900–2000: A Centennial History* (Pittsburgh: Carnegie Mellon University Press, 2000); Roy Lubove, *Twentieth-Century Pittsburgh*; vol. 1, *Government, Business and Environmental Change* (Pittsburgh: University of Pittsburgh Press, 1995), 119.

13. Interview with Madoff, Surprise Arizona, October 15–16, 2007.

14. Quote from telephone interview with Madoff, February 14, 2002; cf. interview with Madoff, Surprise, Arizona, October 15–16, 2007. Madoff is referring to John P. Hoerr, later the author of, among other nationally known works, *And the Wolf Finally Came: The Decline of the American Steel Industry* (Pittsburgh: University of Pittsburgh Press, 1988). Madoff provides a punchline: "That's how he was created and that's how I was created."

15. For more on the postwar suburbanization of Pittsburgh's working class, see Kent James, "Public Policy and the Postwar Suburbanization of Pittsburgh, 1945–1990" (Ph.D. dissertation, Carnegie Mellon University, 2005); Joel A. Tarr and Denise DiPasquale, "The Mill Town in the Industrial City: Pittsburgh's Hazelwood," *Urbanism Past & Present* 7, 1 (1982): 1–14; for GASP and Squirrel Hill, see GASP Board of Directors' meeting, February 23, 1970, ff "GASP—Records—early material 1969–1970," HSWP GASP. For more on Squirrel Hill, see James L. Longhurst, "'Don't Hold Your Breath, Fight for It!' Women's Activism and Citizen Standing in Pittsburgh and the United States, 1965–1975" (Ph.D. diss., Carnegie Mellon University, 2004), 163.

16. "Ethnic Groups 1971," ff "Public Relations Mailing Lists," GASP Internal. While GASP included ethnic groups on mailing lists, neither they nor any African American organizations were counted as a part of GASP's "representative membership." See Table 3.1.

17. Robert D. Putnam, *Bowling Alone: The Collapse and Revival of American Community* (New York: Simon & Schuster, 2000), 19. For more on this topic, see the discussion of social capital in the Preface.

18. "New Talks Set," *PPG* 12/2/69; "Grape Boycott March," *PPG* 3/1/70. Protests against

Gulf Oil turned violent later in the decade, as the Weather Underground detonated a bomb in the company's Grant Street corporate headquarters; but that action was unrelated to local protest. "Gulf Bldg. Ripped by Dynamite," *PPG* 6/14/74; NOW reference from Eleanor Humes Haney, *A Feminist Legacy: The Ethics of Wilma Scott Heide and Company* (Buffalo, NY: Margaretdaughters, 1985), 59–62; quote from Bloom telephone interviews with author, September 12, 2007, and February 26, 2009.

19. "By-Laws, 1970 Old" in ff "By-Laws," GASP Internal; "By-laws, dated 5/18/1970," ff "GASP BOARD & planning committee," AIS Pelkofer.

20. "Clean Air Crusaders," *PP* 11/30/69; "GASP Praises 4," *PP* 10/24/70; "Top Polluters," *PP* 10/27/70.

21. "Hunt Has Doubts," *PP* 11/2/69; letters to the editor, *PP* 11/11 and 11/12/69; "Hunt Air Curb Criticized," *PP* 11/12/69; "Clearing the Air" editorial, *PP* 11/14/69.

22. "Don't Hold Your Breath . . . Fight For It!" n.d., 2, CLP GASP; untitled GASP pamphlet, n.d. (ca. 1970), 1, CLP GASP; *A Breath of Fresh Air*, audio recording of Voice of America documentary, CLP GASP.

23. Fisher, *Let the People Decide*, 140.

24. A claim of 30,000 appeared in "Sixth Grade Pupils Gasp," *PP* 3/26/70; 40,000 was claimed in "Don't Hold Your Breath . . . Fight For It!" The 60,000 estimate comes from "Dirtie-Gertie," *Squirrel Hill News* 3/4/71, and also untitled GASP pamphlet, n.d., 2, CLP GASP, as well as Arthur Kitzes to George Schultz, Office of Management and Budget, Executive Office of the President, Washington DC, December 7, 1971, ff "Early GASP Correspondence RE: Clean Air Act," AIS Pelkofer. The intention of reaching 100,000 members is listed in "GASP—WHO we are, WHAT our views are and HOW we operate: . . . Madoff, . . . April 22nd, 1970." Typescript dated 5/1/70, AIS Pelkofer.

25. Jones, *Clean Air*, 151; Helen Frank Collins, "Between Gasps of Fresh Air . . ." *The Christian Science Monitor* (February 20, 1970): 2.

26. Internal Revenue Service Form 990 for 1971 and 1972, ff "Income Tax," GASP Internal; Membership dues for 1971–1972 from pamphlet, "Don't Hold Your Breath . . . Fight For It!" ff "Membership Forms," GASP Internal.

27. Interview with Madoff, Surprise, Arizona, October 15–16, 2007.

28. Compiled from five separate lists: "Rep 8/73," Janet Blystone to Representative Membership groups, n.d., ca 1971–1973, "Representative Membership, n.d.," unnamed list of representative membership, and list of organizations involved in cookie project, dated October 17, 1971, and February 1971, all from ff "Representative Membership," GASP Internal. Total number of organizations listed was 588, total unique was 266.

29. Ff "Representative Membership," "Rep. 8/73."

30. Ibid.

31. "Transcripts of Hearings on Proposed Regulations, 1969, Written Statements," PSA RG 11.43.

32. Putnam, *Bowling Alone*, 19.

33. Baird Straughan and Tom Pollak, "The Broader Movement: Nonprofit Environmental and Conservation Organizations, 1989–2005" (Washington, DC: National Center for Charitable Statistics at the Urban Institute, 2008), 1; for examples, see Marc Mowrey and Tim Redmond, *Not in Our Backyard: The People and Events That Shaped America's Modern Environmental Movement* (New York: William Morrow, 1993), 13; Michael E. Kraft, "U.S. Environmental Policy and Politics: From the 1960s to the 1990s," *Journal of Policy History* (2000): 23; and discussion in the Preface of this book.

34. There is dispute on this point. See Putnam, *Bowling Alone*, 160, n. 54.

35. Yearly extracts from the Internal Revenue Service "Business Master File," all twentieth-century founded organizations granted 501(c)(3) status, Master-05/2006 from the National Center for Charitable Statistics, http://nccsweb.org. Some caveats: while the IRS eliminates defunct organizations from its yearly BMF, the NCCS has kept defunct organizations since 1989, thus underrepresenting older organizations and partially explaining the large spike toward the end of the century. The 501(c)(3) label was created in 1954, and this data uses equivalent categories from before both the 1939 code and its 1954 revision. Cf. Straughan and Pollak, "The Broader Movement," 12; Nicholas Freudenberg and Carol Steinsapir, "Not in Our Backyards: The Grassroots Environmental Movement" in *American Environmentalism: The U.S. Environmental Movement, 1970–1990*, ed. Riley Dunlap and Angela Mertig (Philadelphia: Taylor & Francis, 1992), 27–38.

36. While the IRS figures demonstrate an increase in all charitable organizations, environmental organizations, and those groups independent from a national body, there are significant limitations to using the IRS figures as representative of all organizing. Not all organizations existed long enough to apply for tax-exempt status. This source underrepresents organizations that ceased to operate before 1989, and entirely misses charitable organizations whose life span ended before the creation of the Internal Revenue Code of 1939.

37. Records of the Department of State, Commission on Charitable Organizations, Applications for Certificate of Registration—Charitable Organizations, PSA RG 26.

38. "Registered Nonprofit Organizations by IRS Ruling Date," NCCSDataweb extract from the January 2007 IRS BMF for Pennsylvania.

39. Andrew W. Martin, Frank R. Baumgartner, and John D. McCarthy, "Measuring Association Populations Using the *Encyclopedia of Associations*: Evidence from the Field of Labor Unions," *Social Science Research* (September 2006): 771–778.

40. *Encyclopedia of Associations* data presented here was accessed September 6, 2007 from the "Associations Unlimited" database from Gale, searching national, regional, state, and local subsets and excluding international organizations.

41. Straughan and Pollak, "The Broader Movement," n. 5.

42. Jones, *Clean Air*, 110–112; Scott H. Dewey, *Don't Breathe the Air: Air Pollution and U.S. Environmental Politics, 1945–1970* (College Station, TX: Texas A&M University Press, 2000), 244; Shawn Bernstein, "The Rise of Air Pollution Control as a National Political Issue" (Ph.D. dissertation, Columbia University, 1986).

43. NAPCA's mission in ff "Briefing Statement on the NAPCA, November 1968," Office of the Administrator, Issuances of Predecessor Agencies, 1966–1971, NARA RG 412.

44. Compilation of ff "OC Citizen's Committee 1971" (1) and (2), "Office of Air Programs Administrative Subject Files, 1966–1972," NARA RG 412.

45. Leighton A. Price, NAPCA Office of Education and Information, to Sheldon Samuels, November 19, 1970, ff "Citizens Committee (2)," NARA RG 412.

46. For brief histories of the EPA, see Edmund P. Russell III, "Lost among the Parts Per Billion: Ecological Protection at the United States Environmental Protection Agency, 1970–1993," *Environmental History* (1997): 29–51; Joel A. Mintz, *Enforcement at the* EPA: High Stakes and Hard Choices (Austin: University of Texas Press, 1995); Marc K. Landy, Marc J. Roberts, and Stephen R. Thomas, *The Environmental Protection Agency: Asking the Wrong Questions* (Oxford: Oxford University Press, 1990); Richard N. L. Andrews, *Managing the Environment, Managing Ourselves: A History of American Environmental Policy* (New Haven: Yale University Press, 1999) 227–237, especially 234; Benjamin Kline, *First along the River: A Brief History of the U.S. Environmental Movement* (San Francisco: Acada Books, 2000), 92–96; Hal K. Roth-

man, *The Greening of a Nation? Environmentalism in the United States Since 1945* (Fort Worth, TX: Harcourt Brace College, 1998), 117–121.

47. Rothman, *The Greening of a Nation*, 117; for a partial list of affected federal programs, see Kline, *First Along the River*, 92; cf. Andrews, *Managing the Environment, Managing Ourselves*, particularly chapter 12; Philip Shabecoff, *A Fierce Green Fire: The American Environmental Movement* (New York: Hill and Wang, 1993), 129–133; Russell, "Lost among the Parts Per Billion," 29–51.

48. See, for example, Loretta A. [illegible], Northside Environmental Action Committee, (Indianapolis, Indiana), to Ruckelshaus, January 2, 1972, ff "State Air Standards," Office of Air Programs Administrative Subject Files, 1966–1972, NARA RG 412.

49. "Citizen Action Can Get Results" (August 1972), U.S. GPO 1972 0-469-955, 8; cf. "The Citizen's Role in Environmental Decision Making" (November 1972), U.S. GPO 1972 0-478-748; GASP commentary on latter in GASP *Hotline* 3 (March–April 1973): 7.

50. Pat Newman to Ruckelshaus, May 5, 1972, ff "EPA 70–75," GASP Internal.

51. "Pollution Fight across Nation," *NYT* 2/24/70.

52. Citizen's Clean Air Committee from Fred M. Nathanson and Gwen Gwinn, "This Industry Cleaned Up the Air—With a Little Prodding," *American Lung Association Bulletin* (March 1975): 10–13; Citizens Against Pollution and Citizens Council for Clean Air from ff "Citizens Committee," Office of Air Programs Administrative Subject Files, 1966–1972, NARA RG 412; Community Action to Reduce Pollution from Andrew Hurley's *Environmental Inequalities: Class, Race and Industrial Pollution in Gary, Indiana, 1945–1980* (Chapel Hill: University of North Carolina Press, 1995); Citizen's Air Conservation Association from "Meeting Packets, 1961–1971," and "Stenographic report of hearing held at State Office Building, . . . April 16, 1969," 103, ff "Transcript of Hearings on Proposed Regulations, 1969, Regulations III, IV, VI," PSA RG 11.42; cf. ff "State Air Standards," NARA RG 412.

53. NCCS Data Web, Master-05/2006.

54. "GASP to Air Smog Plan," *PP* 3/28/71.

55. Lowella H. "Bunny" Butler to Ruckelshaus, El Paso, TX, March 24, 1972, ff "State Air Standards," NARA RG 412; several letters, first dated September 15, 1969, from the "Greater-Washington Alliance to Stop Pollution, Inc." ff "OC-1," NARA RG 412; El Paso (TX) *Herald-Post* (September 29, 1971); Dewey, *Don't Breathe the Air*, 106. None of these GASPs should be confused with local chapters of the national Group Against Smoking Pollution. See Judith Layzer, *The Environmental Case: Translating Values into Policy*, 2nd ed., (Washington, DC: CQ Press, 2006), 31. Cf. Frank Uekoetter, *The Age of Smoke: Environmental Policy in Germany and the United States, 1880–1970* (Pittsburgh: University of Pittsburgh Press, 2009), 222–223.

56. Interview with Madoff, Surprise, Arizona, October 15–16, 2007.

57. "GASP at 10," *PPG* 12/4/79; "How Bill Rod Saw 1973," *PF* 1/4/74.

58. On NIMBY, see Michael E. Kraft and Bruce B. Clary, "Citizen Participation and the Nimby Syndrome: Public Response to Radioactive Waste Disposal," *Western Political Quarterly* 44 (1991): 299–328; Bo Poertner, "Is Latest Criticism Worthwhile Talk or Just Worthless?" Orlando (FL) *Sentinel* (September 30, 1990); for more on the relationship between NIMBY protest and environmental justice organizing, see Eileen Maura McGurty, "From Nimby to Civil Rights: The Origins of the Environmental Justice Movement," *Environmental History* 2 (1997): 301–323; Martin V. Melosi, "Environmental Justice, Political Agenda Setting, and the Myths of History," *Journal of Policy History* 12 (2000): 43–71.

59. For CSE see "Think Tanks: Corporations' Quiet Weapon," *Washington Post* (January 29, 2000); for COMPASS see Sharon Beder, *Global Spin: The Corporate Assault on Environmental-*

ism (White River Junction, VT: Chelsea Green, 2002), 244, also 27–45, 122; for SOSA see Paul D. Thacker, "Hidden Ties," *Environmental Science and Technology* (May 15, 2006): 3128–3134. On the Astroturf phenomenon in general see Thomas P. Lyon, "Astroturf: Interest Group Lobbying and Corporate Strategy," *Journal of Economics and Management Strategy* (2004): 561; John McNutt and Katherine Boland, "Astroturf, Technology and the Future of Community Mobilization: Implications for Nonprofit Theory," *Journal of Sociology and Social Welfare* (September 2007): 165–178; "A New Breed . . ." *NYT* 3/17/93; quote from Michael M. Gunter, Jr., "Review of Sharon Beder, *Global Spin*," *Organization and Environment* (2003): 402.

60. For ROAR see *Newsweek* (June 9, 1975): 31; Bruce Stutz, "Britain's Elusive Eco-Town Dream," *On Earth* 30 (Winter 2009): 26–33.

61. MOBE, pronounced "Moby," was not an acronym, but a nickname for The National Mobilization Committee to End the War in Vietnam, previously the Spring Mobilization Committee to End the War in Vietnam; CORE was and is the Congress on Racial Equality; for more on SNCC, ACORN, and PUSH see Chapter 1.

62. Clarence Lang, "Between Civil Rights and Black Power in the Gateway City: The Action Committee to Improve Opportunities for Negroes (ACTION), 1964–75," *Journal of Social History* 37 (Spring 2004): 725–754.

63. Out of 11,402 total names post-1960 coded "Environmental," NCCS Data Web, Master-05/2006; STOP is in Schudson, *The Good Citizen*, 290.

64. GOO in Ted Steinberg's *Down to Earth: Nature's Role in American History*, 2nd ed. (Oxford: Oxford University Press, 2009), 249; *United States et al. v. Students Challenging Regulatory Agency Procedures* (SCRAP) et al., 412 U.S. 669 (1972); Lois Marie Gibbs, *Love Canal: The Story Continues* (Gabriola Island, BC: New Society, 1998); Allan Mazur, *A Hazardous Inquiry: The Rashomon Effect at Love Canal* (Cambridge, MA: Harvard University Press, 1998); Elizabeth Blum, *Love Canal Revisited: Race, Class, and Gender in Environmental Activism* (Lawrence: University Press of Kansas, 2008).

65. NCCS Data Web Master-05/2006.

66. Schudson, *Good Citizen*, 279–280; Robert D. Putnam, "Bowling Alone: America's Declining Social Capital," *Journal of Democracy* 6 (1995): 67.

Chapter 4: Mothers of Urban Skies

1. Jeannette Widom, "Party Cookies Only," GASP pamphlet printed 1972, vi–vii. The text quoted here appeared in multiple GASP publications.

2. Charles O. Jones, *Clean Air: The Policies and Politics of Pollution Control* (Pittsburgh: University of Pittsburgh Press, 1975); Robert Gottlieb, *Forcing the Spring: The Transformation of the American Environmental Movement* (Washington, DC: Island Press, 1993), 126–127.

3. Maureen A. Flanagan, *Seeing with Their Hearts: Chicago Women and the Vision of the Good City, 1871–1933* (Princeton: Princeton University Press, 2002); Suellen M. Hoy, "'Municipal Housekeeping': The Role of Women in Improving Urban Sanitation Practices," in *Pollution and Reform in American Cities, 1870–1930*, ed. Martin V. Melosi (Austin: University of Texas Press, 1980), 173–198; Andrea Tuttle Kornbluh, *Lighting the Way: The Woman's City Club of Cincinnati, 1915–1965* (Cincinnati, OH: Young & Klein, 1986), 1.

4. Virginia Scharff, ed., *Seeing Nature through Gender* (Lawrence: University Press of Kansas, 2003); see particularly the work of Carolyn Merchant, perhaps beginning with *Earthcare: Women and the Environment* (New York: Routledge, 1995); Adam Rome, "Give Earth a Chance: The Environmental Movement and the Sixties," *Journal of American History* 90.2 (2003): 525–554.

5. For accounts of women's leadership in the postwar environmental movement, see Gottlieb, *Forcing the Spring*; Allan Mazur, *A Hazardous Inquiry: The Rashomon Effect at Love Canal* (Cambridge, MA: Harvard University Press, 1998); Jack E. Davis, "'Conservation Is Now a Dead Word': Marjory Stoneman Douglas and the Transformation of American Environmentalism," *Environmental History* 8 (2003): 53–76; Lois Marie Gibbs, *Love Canal: The Story Continues* (Gabriola Island, BC: New Society, 1998); Elizabeth Blum, *Love Canal Revisited: Race, Class, and Gender at Love Canal* (Lawrence: University Press of Kansas, 2008), especially chapter 2 n. 5; Linda J. Lear, *Rachel Carson: Witness for Nature* (New York: Henry Holt, 1997).

6. Sherie R. Mershon, "Corporate Social Responsibility and Urban Revitalization: An Organizational History of the Allegheny Conference on Community Development, 1943–1990" (Ph.D. dissertation, Carnegie Mellon University, 2000).

7. Mershon, "Corporate Social Responsibility," 3, 297.

8. Michelle Madoff to "Fellow Breathers," October 1969, AIS Madoff; "GASP Board of director's meeting Feb 23, 1970," ff "GASP-Records—early material—1969–1970," HSWP GASP.

9. The first six GASP presidents were Michelle Madoff, 1969–summer 1972; Ann Cardinal, fall 1972; Pat Newman, 1972–1973; Arnold Kitzes, 1973–1974; Arthur Gorr, 1974–1975; Joan Hays, fall 1975. See James L. Longhurst, "'Don't Hold Your Breath, Fight for It!' Women's Activism and Citizen Standing in Pittsburgh and the United States, 1965–1975" (Ph.D. diss., Carnegie Mellon University, 2004), Table 3.1; ff "Letterheads—Historical," "Correspondence I," "Correspondence II," "Correspondence—Pennsylvania Legislature," "Correspondence—Gorr," "Correspondence 74," "Correspondence 75," all from GASP Internal.

10. Compilation of attendance at all archived minutes of Board of Directors meetings, February 1970–December 1975. Not all minutes include named attendance. CLP GASP; AIS Pelkofer; AIS Madoff; GASP Internal.

11. For general information on Madoff, see Madoff interview with author, Surprise, Arizona, October 15–16, 2007; "Michelle Madoff," ff "GASP Staff Bios," GASP Internal; Charles O. Jones, *Clean Air: The Policies and Politics of Pollution Control* (Pittsburgh: University of Pittsburgh Press, 1975), 149–151; John O. Frohliger, "GASP: An Ecological David," *Communiqué: Journal of the University of Pittsburgh Hospitals* (1972): 12–16. Madoff resignation from GASP Board in Madoff to Ann Cardinal, October 10, 1972, CLP GASP. On Madoff's political career after GASP, see "Michelle Madoff Named to State Advisory Post" GASP *Hotline* (May 1971): 3; "Independent Michelle Madoff," *PPG* 5/26/75; quotes from "Sixth Grade Pupils GASP," *PP* 3/26/70, "Crusading Lady a Pittsburgh Symbol," Rochester (NY) *Democrat Chronicle* (October 26, 1971), also in "Impressions of a Changing City," *PPG* 4/14/74.

12. "Marilyn F. Janocko," ff "GASP Staff Bios," GASP Internal; letter to the editor, *PPG* 7/31/74; "Not Easy Being an Environmentalist," *PPG* 7/31/82.

13. "GASP Leader Urges," Clairton *Bulletin-Times* September 1,1971; letter to the editor, *PPG* 11/26/71; Madoff interview with author, Surprise, Arizona, October 15–16, 2007.

14. "Citizens Wising Up," *PP* 11/26/72; "EPA Region III: 20 Years of Making a Difference," EPA publication # 903R90003 (December 1990), 26.

15. "Women in the Environment," *PF* 11/24/72; "Remembering Patricia Pelkofer," *PPG* 5/29/03.

16. Interview with the author, Pittsburgh, October 1, 2002; Sylvia Sachs, "Works of Art: Cookie Baker Virtuoso," *PP*, n. d., ff "Clippings I," GASP Internal; "Cooking Champ," *South Hills Shopping Guide* (September 8, 1973): 8; "After 25 Years in the 'Stenches,'" *PP* 11/23/71; "Cookies a la Machine," *PPG* 12/12/96; AIS Widom.

17. "Pittsburghers Top Cleanup Fighters," *PP* 2/1/73; "Women in the Environment," *PF* 11/24/72.

18. GASP pamphlet "She's Some Cookie! Dirtie Gertie the Poor Polluted Birdie—A Capsule History," ff "Cookbooks—PCO I," GASP Internal; "'Dirtie Gertie' Bake Sale Arcade Feature," *Squirrel Hill News* 3/4/71; "'Dirtie Gertie' Cookies!" *East End Tribune* (March 3, 1971; GASP press releases, February 1971, October 17, 1971, November 17, 1973, ff "Representative Membership," GASP Internal; ff "Promotions, 1970–1971," AIS Widom.

19. GASP pamphlet "She's Some Cookie!"; "GASP Celebrates Birthday," Penn Hills *Progress* December 4, 1974, 15; "Special Projects Chairman's Report: Fighting Pollution with a Rolling Pin," GASP *Hotline* 3 (February 1973): 10.

20. "City GASPing for Canned Air," *PP* 1/16/70; film, *I Belong Here!*; "Air Pollution Emergency," *PP* 1/16/70; "Pollution Fighters," *PP* 2/19/71.

21. Jeannette Widom to Mrs. Gilliam, July 15, 1974, ff "Cookbooks—Just Coffeecake," GASP Internal; "Works of Art: Cookie Baker Virtuoso," *PP* n.d., in author's possession.

22. "Works of Art"; Jeannette Widom to Arlene Taylor, July 2, 1974, ff "Cookbooks—Just Coffeecake," GASP Internal.

23. Longhurst, "Don't Hold Your Breath," 183–190; "Return of Organization Exempt from Income Tax," IRS Form 990, 1972, ff "Income Tax," GASP Internal; "Monthly Gross Sales . . ." October 1, 1975, ff. "Cookbooks—PCO I," GASP Internal.

24. "Sponsor's List," ff "Cookbooks—Just Coffeecake," GASP Internal; "Fun Buns Press Party Guest List," ff "Fun Buns," GASP Internal.

25. Jeannette Widom, "Fun Buns for Kids to Make, Bake, Decorate and Eat" (1975), 41–46, AIS Widom. For further discussion of the county's Air Pollution Index and GASP's role in promoting it, see James Longhurst, "1 to 100: Creating an Air Quality Index in Pittsburgh," *Environmental Monitoring and Assessment* 106 (July 2005): 27–45.

26. "GASP Cookbooks: Sweeten your baking—and the air too!" ff "Fun Buns"; "Party Cookies Only: A Specialty Cookbook Presented by GASP," ff "Cookbooks—PCO I"; "GASP Announces 'Fun Buns for Kids' Project to Raise Dough for Clean Air," press release, December 3, 1975, ff "Fun Buns." All from GASP Internal.

27. Esther G. Kitzes was married to GASP member Dr. Arnold S. Kitzes, biographical information from Kitzes to Office of Environmental Education, Washington, DC, April 18, 1972, enclosure grant proposal "Project Environmental Rebirth (PER) Phase II, 7, ff "Proposals—DHYB, PER," GASP Internal. Kitzes passed away in March 1973, before the completion of the second GASP film, which was dedicated to her. GASP *Hotline* (May 1973): 2; obituary, *PPG* 3/24/73.

28. "Proposal for Project Grant Under Public Law 91-516," 15, ff "Proposals—DHYB, PER," GASP Internal.

29. The $72,200 grant proposal was turned down. Dr. Arnold S. Kitzes to Office of Environmental Education, U.S. Office of Education, Washington, DC, April 18, 1972, ff "Proposals—DHYB, PER," GASP Internal; IRS Form 990 for 1971, ff "Income Tax," GASP Internal; GASP *Hotline* 2 (March–April 1972): 1.

30. Although twenty copies of this film were made to be distributed and offered for rental, I know of only a single extant copy: *Don't Hold Your Breath (Fight For It!)*, AIS Pelkofer.

31. GASP *Hotline* 2 (March–April 1972): 6.

32. "'Dirty Movie' Features Pollution," *PP* 2/16/72, 24.

33. GASP press release, "I Belong Here: A New Film Entertainment on Citizen Involvement in Environmental Action," January 1975, ff "Films," GASP Internal.

34. Film, *I Belong Here!*

35. Receipts and correspondence, ff "Don't Hold Your Breath," GASP Internal.

36. "'Clean Air' Comes in Flip-Top Cans," Braintree (MA) *Observer* (September 24, 1971).

37. GASP *Hotline* 2 (November 1972): 3.

38. GASP Board of Directors' Meeting Minutes, November 3, 1973, 2–3, CLP GASP; "GASP Leadership Seminar for Citizen Action: Participant Profiles," ff "Leadership Seminar," GASP Internal.

39. "Dirtie Gertie: Be My Valentine," GASP *Hotline* 2 (March–April 1972): 5.

40. "Pollution Land Tour," ff "Pollution Land Tour," GASP Internal; "GASP Tour," *PF* 3/12/71.

41. "Earth Week," *PP* 4/21/71; press release, June 14, 1972, ff "PR 1972"; folders "PER I," "PER II," "PER III," all from GASP Internal; "Students Attend Pilot Environmental Program," *Progress* (August 23, 1972): 22.

42. "Brochure," ff "Correspondence," GASP Internal; "GASP Announces 'Fun Buns for Kids' Project to Raise Dough for Clean Air," press release (December 3, 1975), 2, ff "Fun Buns," GASP Internal; film, *Don't Hold Your Breath*; "GASP Issues Alert," *PPG* 12/3/71; "Sixth Grade Pupils," *PP* 3/26/70; "Car-Free," *PP* 1/26/71. The image of mothers marching with strollers is particularly powerful symbolism; see Amy Swerdlow, *Women Strike for Peace: Traditional Motherhood and Radical Politics in the 1960s* (Chicago: University of Chicago Press, 1993), 2–3.

43. Quotes from "Don't Hold Your Breath . . . Fight For It!" GASP pamphlet, n.d. (ca. 1970), 2, CLP GASP.

44. Unsigned letter to "Ruth Heimbuecher, Features Dept, Editorial Room, PPress," January 20, n.d, CLP GASP; cf. GASP press release, April 6, 1972, ff "PR 1972," GASP Internal.

45. "Dirtie Gertie's Baby Contest," GASP *Hotline* (March–April 1972): 10, AIS Pelkofer.

46. Jeannette Widom, "Fun Buns," page 42–43. Question mark in parentheses is in original.

47. Untitled Dirtie Gertie script, ff "Dirtie Gertie," GASP Internal.

48. "Dirtie Gertie Now a 'Good Scout'" GASP *Hotline* (June 1972): 5.

49. Vera Norwood, *Made from This Earth: American Women and Nature* (Chapel Hill: University of North Carolina Press, 1993), 148.

50. Vera Norwood, "Women's Roles in Nature Study and Environmental Protection," *OAH Magazine of History* 10 (1996): 17.

51. "Clean Air Crusaders Start Breathing Fire," *PP* 11/3/69; Helen Frank Collins, "Between Gasps of Fresh Air . . ." *Christian Science Monitor* (February 20, 1970): 2; "Wife Wars against Pollution," *PP* 12/2/71.

52. "GASP Leader Urges Get Involved, Act, Grow!" Clairton *Bulletin-Times* September 1, 1971. For a contemporaneous discussion of DDT in mother's milk, see Rachel Carson, *Silent Spring* (Cambridge, MA: Houghton Mifflin, 1962), 23; cf. Norwood, *Made from This Earth*, 155.

53. "Clean Air Warriors," *PPG* 10/20/70; "Will Anti-Pollution Laws Stick?" *WSJ* 6/2/70; "State's Clean Air Standards," *PPG* 11/10/71; "Pittsburgh Cracks Down on Polluters," *Business Week* (December 27, 1969); "Sixth Grade Pupils," *PP* 3/26/70; "GASP Founder, Hunt," Donora [Monongahela, PA] *Daily Herald* 7/26/72.

54. Pat Newman, "A Report on an Address by William Tipton . . .," ff "EPA 70-75," GASP Internal; Jones, *Clean Air*, 151; Madoff letter to the editor, *PPG*, 6/3/74; cf. Scott H. Dewey, *Don't Breathe the Air: Air Pollution and U.S. Environmental Politics, 1945–1970* (College Station: Texas A&M University Press, 2000), 153–4.

55. "Stenographic report of hearing . . . December 4, 1971," 635, PSA RG 43:16; "Who Are the Environmentalists?" GASP *Hotline* (October 1973): 5; Madoff's and GASP's fight against the labels of hysteria and emotionalism echoed the criticism of earlier female activists, including Rachel Carson, whom the science editor for *Time* magazine once described as "hysterically overemphatic" and her supporters as "faddists and hysterical women." Norwood, *Made from This Earth*, 168; c.f. "Women in the Environment," *PF* 11/24/72.

56. "About GASP," n.d., AIS Pelkofer.

57. *A Breath of Fresh Air*, audio recording of Voice of America documentary, CLP GASP.

58. "Something Rotten," *PP* 7/21/72; "Hunt Dismisses," *PPG* 10/17/75; "Madoff Gloats," *PPG* 11/6/75; Carnegie-Mellon University *Tartan* (March 11, 1975); "Candidate Likes," *PPG* 10/27/75.

59. Collins, "Between Gasps of Fresh Air . . .,"2. Emphasis in original.

60. "Stenographic Report of hearing . . . January 17, 1972," ff "Implementation Plan Hearing January 17, 1972—Pittsburgh," PSA RG 43.36; GASP *Hotline*,(July 1971): 2; "Clean Air Warriors," *PPG* 10/20/70, "Low Profile Organization," *PPG* 3/3/73; Albany (NY) *Times-Union* (October 23, 1971).

61. "Stenographic report of hearing . . . Pittsburgh, Pennsylvania, Friday December 3, 1971, at 9 a.m.," 461, PSA RG 43.16.

62. "Air Pollution Hearing: Pittsburgh Pennsylvania, Tuesday September 9, 1969," 210–211, PSA RG 11.43.

63. "Concerned Mother," *PP* 9/9/69; "50 Air Strong Views," *PPG* 12/4/71.

64. Telephone interview with Madoff, February 14, 2002; interview, October 15, 2007, Surprise, Arizona.

65. Interview with Madoff, October 15, 2007; "Ms. Madoff Files Suit," *PP* 10/24/75; "Madoff Files for Divorce," *PPG* 1/23/81.

66. Robyn Muncy, *Creating a Female Dominion in American Reform, 1890–1935* (New York: Oxford University Press, 1991), xv; Theda Skocpol, *Protecting Soldiers and Mothers: The Political Origins of Social Policy in the United States* (Cambridge, MA: Belknap, 1992), first quote at 3, see also 20–21; second quote from 368, see also 371; Seth Koven and Sonya Michel, eds., *Mothers of a New World: Maternalist Politics and the Origins of Welfare States* (New York: Routledge, 1993), 29.

67. See, among many, Elizabeth J. Clapp, *Mothers of All Children: Women Reformers and the Rise of Juvenile Courts in Progressive Era America* (University Park: Pennsylvania State University Press, 1998); Molly Ladd-Taylor, "Toward Defining Maternalism in U.S. History," *Journal of Women's History* 6 (1993): 110–113; Seth Koven and Sonya Michel, "Womanly Duties: Maternalist Politics and the Origins of Welfare States in France, Germany, Great Britain, and the United States, 1880–1920," *American Historical Review* 95 (1990): 1076–1108; Sonya Michel and Robyn Rosen, "The Paradox of Maternalism: Elizabeth Lowell Putnam and the American Welfare State," *Gender and History* 4 (1992): 364–386; and essays by Judith Walzer Leavitt, Linda Gordon, and Estelle B. Freedman in *U.S History as Women's History: New Feminist Essays*, ed. Linda K. Kerber, Alice Kessler-Harris, and Kathryn Kish Sklar (Chapel Hill: University of North Carolina Press, 1995).

68. Carolyn Merchant, "Women of the Progressive Conservation Movement: 1900–1916," *Environmental Review* 8 (1984): quote from 57, see also 74 and 80; Judith W. Leavitt, *The Healthiest City: Milwaukee and the Politics of Health Reform* (Princeton: Princeton University Press), 1982; Hoy, "Municipal Housekeeping," 194.

69. Quote from Maureen A. Flanagan, "The City Profitable, the City Livable," *Journal of Urban History* 22 (1996): 183. See also Flanagan, "Gender and Urban Political Reform: the City Club and the Women's City Club of Chicago in the Progressive Era," *American Historical Review* 95 (1990): 1032–1050; and Flanagan, *Seeing with their Hearts*, 5; Angela Gugliotta, "Class, Gender, and Coal Smoke: Gender Ideology and Environmental Justice in Pittsburgh, 1868–1914," *Environmental History* 5 (2000): 165–197.

70. Susan Lynn, *Progressive Women in Conservative Times: Racial Justice, Peace, and Feminism, 1945–1960s* (New Brunswick, NJ: Rutgers University Press, 1992), 3.

71. Swerdlow, *Women Strike for Peace*, 2–3.

72. See "Air Pollution: Clubwomen Accept the Challenge," address by Thomas F. Williams, presented at the 75th annual meeting of the GFWC June 9, 1966, U.S. GPO FS 2.30: W 67/4, 15270 (1966).

Chapter 5: "Where the Rubber Meets the Road"

1. "Conservation Group Rips," *PP* 11/23/70; "U.S. Steel Asks," *PP* 6/30/70; "Power Firm Chided," *PPG* 1/23/70; "County 'No' Expected," *PP* 3/25/72; "County Fumes at CMU," *PP* 1/30/73.

2. Marc K. Landy, Marc J. Roberts, and Stephen R. Thomas, *The Environmental Protection Agency: Asking the Wrong Questions* (Oxford: Oxford University Press, 1990), 204; cf. Joel A. Mintz, *Enforcement at the EPA: High Stakes and Hard Choices* (Austin: University of Texas Press, 1995), 10. See also Mary J. Coulombe, "Exercising the Right to Object: A Brief History of the Forest Service Appeals Process," *Journal of Forestry* 102 (March 2004): 10–13.

3. James Pearson, *Reducing the Averages: The Founding and Development of the Puget Sound Air Pollution Control Agency* (San Jose, CA: Writer's Club Press, 1999).

4. Allegheny County Health Department Rules and Regulations, Article XVII, January 1, 1970.

5. "Pittsburgh Cracks Down," *Business Week* (December 27, 1969): 67; "New Air Pollution Rules," *PPG* 12/4/69; see also Nicholas Casner, "Acid Mine Drainage and Pittsburgh's Water Quality," in *Devastation and Renewal: An Environmental History of Pittsburgh and Its Region*, ed. Joel A. Tarr (Pittsburgh: University of Pittsburgh Press, 2003), 89–109.

6. Donald P. Eriksen, "Air Pollution: The Problem and the Legislative and Administrative Responses of the United States, Pennsylvania and Allegheny County," *University of Pittsburgh Law Review* 30 (1968): 633–670; ACHD Annual Reports, 1966–76, in the collection of the Falk Medical Library, University of Pittsburgh; James L. Longhurst, "'Don't Hold Your Breath, Fight for It!' Women's Activism and Citizen Standing in Pittsburgh and the United States, 1965–1975" (Ph.D. diss., Carnegie Mellon University, 2004), Table 5.1.

7. For more on coke quenching see Chapter 6, and ACHD "Annual Report" (1969), 7; Article XVII (January 1, 1970) § 1708.2; Robert W. Dunlap and Michael J. Massey, "Assessment of Alternatives," May 29, 1971, ff "Variance Board—Technology," GASP Internal.

8. Article XVII (January 1, 1970) § 1708.2.

9. "Pittsburgh Cracks Down," 67; "GASP Plans Ahead," *PF* 10/30/70.

10. For biographical information see Charles O. Jones, *Clean Air: The Policies and Politics of Pollution Control* (Pittsburgh: University of Pittsburgh Press, 1975), 215. For more on Nickeson, see "Housewife Gets Post," *PP* 8/14/60.

11. "You Can't Tell The Players Without A Program," August 17, 1971, ff "Correspondence," GASP Internal.

12. Cf. Jones, *Clean Air*, 218, table 16; Longhurst, "Don't Hold Your Breath," 238 n. 22.

13. GASP *Hotline* (June 1970); "Ban on Burning," *PP* 1/26/71; "11 Towns Ask County," *PPG* 1/26/71; quote from "2 Official's," *PPG* 5/13/70.

14. "Robert M. Chambers, Inc.," hearing no. 129, docket no. 155, in "Variance Compliance, 1970–1971," ff "Variance Board," GASP Internal.

15. Hearing 80, docket 14, in "Variance Compliance, 1970–1971"; "County Fumes at CMU," *PP* 1/30/73.

16. Jones, *Clean Air*, 219.

17. Ibid.

18. "Air Pollution Ruling Hit," *PPG* 9/2/70; Fein quote from "State Raps USS," *PP* 2/19/71; "Pollution Appeal Hit," *PPG* 2/19/71.

19. Philip J. Ethington, *The Public City: The Political Construction of Urban Life in San Francisco, 1850–1900* (New York: Cambridge University Press, 1994), 15.

20. Jones, *Clean Air*, 219; cf. Ethington, *The Public City*, 29.

21. GASP *Hotline* (July 4, 1971); "GASP Plans Ahead," *PF* 10/30/70; "You Can't Tell The Players Without A Program," letter to GASP Members, August 17, 1971, ff "Correspondence," GASP Internal; GASP *Hotline* (June 15, 1971).

22. GASP *Hotline* (June 15, 1971).

23. Jurgen Habermas, *The Structural Transformation of the Public Sphere: An Inquiry into a Category of Bourgeois Society* (Cambridge, MA: MIT Press, 1989); quotes from Habermas, "The Public Sphere," in *Critical Theory and Society: A Reader*, ed. Stephen Eric Bronner and Douglas MacKay Kellner (New York: Routledge, 1989), 141.

24. "Light Firm Asks Delay," *PPG* 10/14/71; "Power Plant Close Vowed," *PPG* 10/15/71; quote from "U.S. Steel's New Commitment," *PPG* 6/12/74.

25. Article XVII hearing 42, dockets 37 and 38.

26. "Transcript of January 24, 1972 Duquesne Light Company appearance . . ." in ff "Duquesne Light, 1970–1972," AIS Broughton.

27. Ibid.

28. "Plants Request," *PPG* 9/29/70; "County 'No' Expected," *PP* 3/25/72.

29. Quoted in Jones, *Clean Air*, 152.

30. There are extensive records documenting the public debate over state air pollution rules, although state regulations are largely outside of the scope of this book. "60 Request Time," *PP* 11/28/71; "State Weighs Powers," *PP* 12/2/71; "Air Code Hearing," *PP* 12/2/71; "State Chamber's," *PP* 12/3/71; ff "Testimony I–IV," and ff "Testimony Workshops," GASP Internal.

31. GASP *Hotline* (June 1972); GASP press release, May 1972, ff "PR 1972," GASP Internal.

32. "Allegheny County Air Pollution Variance Board Decision, June 11, 1973," docket no 165C 16, AIS GASP; for more on Picadio, see "Lawyer Gets Air Unit Post," *PPG* 2/9/73.

33. Telephone interview with Bernie Bloom, February 26, 2009. Tape and transcript in author's possession.

34. Chleboski to Variance Board, January 30, 1973; Venable to "Public Interest Groups and Other Intervenors of Allegheny County," February 6, 1973; Dunlap to Venable, January 23, 1973; Venable to Dunlap, January 24, 1973; Venable to Chleboski, January 17, 1973, AIS GASP. Dunlap went on to found the interdisciplinary Engineering and Public Policy Program at Carnegie Mellon, lead a string of successful technology firms, and serve as a Carnegie Mellon trustee before his death in 2009. "Robert W. Dunlap, 71 . . ." Wellesley (ME) *Townsman* (March 26, 2009).

35. "DeNardo Backs Duquesne," *PPG* 2/27/73.

36. Jones, *Clean Air*, 216; ff "Task Force Report," GASP Internal; Madoff CETF memo to GASP board, November 1973; CLP GASP; ff "Enviro-SOS Reports," GASP Internal; letter from Madoff to author, July 6, 2009.

37. "Our Home in Mellon Institute," *Air Repair* (November 1951): 1; William G. Christy, "History of the Air Pollution Control Association," *Journal of the Air Pollution Control Association* (April 1960): 126–137.

38. James Longhurst, "Smoky Ol' Town: The Significance of Pittsburgh in U.S. Air Pollution History," *EM: The Journal of the Air and Waste Management Association* (June 2007): 13–15.

39. Morton Corn and Lawrence DeMaio, "Sulfate Particulates: Size Distribution in Pittsburgh Air," *Science* 143 (February 21, 1964): 803–804; Morton Corn and Thomas L. Montgom-

ery, "Atmospheric Particulates: Specific Surface Areas and Densities," *Science* 159 (March 22, 1968): 1350–1351; Phillip Antommaria, Morton Corn, and Lawrence DeMaio, "Airborne Particulates in Pittsburgh," *Science* 150 (December 10, 1965): 1476–1477.

40. The name was changed from Carnegie Institute of Technology in 1967 when the school merged with the Mellon Institute. See Edwin Fenton, *Carnegie Mellon 1900–2000: A Centennial History* (Pittsburgh: Carnegie Mellon University Press, 2000); for Lave and Seskin's work, see Longhurst, "Don't Hold Your Breath," 267 n. 11.

41. Sumner B. Ely, "General Atmospheric Pollution: Some Work in Pittsburgh Air Pollution Problems," *American Journal of Public Health and the Nation's Health* 38 (July 1948): 961; Longhurst, "Don't Hold Your Breath," 268 n. 13; Longhurst, "1 to 100: Creating An Air Quality Index In Pittsburgh," *Environmental Monitoring and Assessment* (2005): 29–30.

42. Nelson W. Hartz, "Instrumentation in Air Pollution," *Air Repair* 2 (November 1952): 45–46.

43. Emerson Venable, "You and the Air You Breathe," *Air Repair* 3 (May 1953): 140; Longhurst, "Don't Hold Your Breath," 269 n. 16.

44. All of the following biographies draw from materials in "GASP Staff Bios," GASP Internal, as well as listed sources. For Lester Lave's contributions to local health debates, see "Health Aid Boost," *PPG* 1/20/71; "Health Costs Higher," *PPG* 6/15/74; "Air Rule Compliance," *PPG* 12/19/77.

45. Sillman letter to the editor, *PPG* 6/14/73.

46. "Soaring Utility Rates," *PPG* 12/7/74; "Utility Study Chief," *PPG* 12/5/75; "Public Hearing," October 18–November 2, 1973, VII, 2979, "Records of the Office of General Enforcement, Technical Records, 1968–1979," NARA RG 412.

47. For representative coverage of Goldburg's contributions to public debates over air pollution matters, see "Power Firm Chided," *PPG* 1/24/70; "County No. 1," *PPG* 4/20/77; "County Air Shows," *PPG* 9/14/78.

48. Esther Kitzes obituary, *PPG* 3/24/73.

49. "Air Pollution Rules," *PPG* 7/10/69; "5 Get Clean Air," *PPG* 3/3/72; "Pitt Professor OK'd," *PPG* 11/12/75; "Worker's Safety Agency," *NYT* 12/20/76.

50. John O. Frohliger, "GASP: An Ecological David," *Communiqué: Journal of the University of Pittsburgh Hospitals* (1972): 12–16; Robert T. Cheng, Morton Corn, and John O. Frohliger, "Contribution to the Reaction Kinetics of Water Soluble Aerosols and SO_2 in Air at PPM Concentrations," *Atmospheric Environment* (December 1971): 987–1008.

51. Letter to the editor, *PPG* 1/15/72; "Resume," ff "GASP Staff Bios," GASP Internal.

52. Jones, *Clean Air*, 215; GASP press release, July 11, 1969, "GASP Newsletters and Bulletins and Earliest Letters," AIS Pelkofer; "'GASP' to Battle Pollution," *Pittsburgh Catholic* 11/14/69; "Ross Man Heads Pollution Appeals," *North Hills News Record*, 4/1/70. Broughton was killed in a mountain-climbing accident in August 1977. "2 Area Climbers," *PPG* 8/3/77; "A Student Remembers," *PPG* 1/5/78; "7 to Receive," *PPG* 2/9/78.

53. Bloom telephone interview, February 26, 2009; Longhurst, "1 to 100," 32–33. For more on Commoner and the Committee for Environmental Information at Washington University, see Michael Egan, *Barry Commoner and the Science of Survival: The Remaking of American Environmentalism* (Cambridge, MA: MIT Press, 2007), chapter 2. Some of Bloom's time in the EPA is recounted in Marc K. Landy, Marc J. Roberts, and Stephen R. Thomas, *The Environmental Protection Agency: Asking the Wrong Questions* (Oxford: Oxford University Press, 1990), 209–221.

54. "Public Meeting," *PP* 11/16/71.

55. GASP *Hotline* (July 4, 1971): 7; GASP *Hotline* (September 1971).

56. Advertisement, *PF* 5/11/71; GASP *Hotline* (September 1971).

57. Bloom telephone interview, September 12, 2007; Madoff interview with author, Surprise, Arizona, October 15–16, 2007.

58. Pat Newman, "Enforcement, Pittsburgh Style," GASP *Hotline* (March 1973).

59. Ibid., 4; Pat Newman, "Non-enforcement and the Double Standard," GASP *Hotline* (July 1973).

60. "66 Signed Up," *PP* 5/2/75; letter to the editor, *PP* 5/3/75.

61. GASP *Hotline* (May 1973), emphasis in original.

Chapter 6: Citizens and the Courts

1. Louis L. Jaffe, "Standing to Secure Judicial Review: Public Actions," *Harvard Law Review* 74 (1961): 1265; for Jaffe's extensive list of publications, see James L. Longhurst, "'Don't Hold Your Breath, Fight for It!' Women's Activism and Citizen Standing in Pittsburgh and the United States, 1965–1975" (Ph.D. diss., Carnegie Mellon University, 2004), 294 n. 3.

2. Louis L. Jaffe, "Standing to Sue in Conservation Suits," September 11, 1969, Airlie House, Warrenton, VA, ff "Conference on Law and the Environment, Program," AIS Broughton.

3. Thomas J. Misa, *A Nation of Steel: The Making of Modern America, 1865–1925* (Johns Hopkins University Press, 1995), ch. 4.

4. Kenneth Warren, *Big Steel: The First Century of the United States Steel Corporation, 1901–2001* (Pittsburgh: University of Pittsburgh Press, 2001), 1, xvii, 41, and table 3.4.

5. John P. Hoerr, *And the Wolf Finally Came: The Decline of the American Steel Industry* (Pittsburgh: University of Pittsburgh Press, 1988), 84–89; Brian Apelt, *The Corporation: A Centennial Biography of the United States Steel Corporation, 1901–2001* (Pittsburgh: Cathedral Publishing, 2000); Apelt, "100 Years of U.S. Steel," *Iron Age New Steel* 17 (2001): 26–28.

6. Warren, *Big Steel*, 101–103; Joel A. Tarr, *The Search for the Ultimate Sink: Urban Pollution in Historical Perspective* (Akron, OH: University of Akron Press, 1996), 394.

7. "U.S. Steel Forced," *NYT* 8/27/72.

8. T. J. Ess, "Jones & Laughlin Pittsburgh Works," *Iron and Steel Engineer* 31 (1954): 76–102; David H. Wollman and Donald H. Inman, *Portraits in Steel: An Illustrated History of Jones & Laughlin Steel Corporation* (Kent, OH: Kent State University Press, 1999), 59, 118.

9. James Longhurst, "Ling, James Joseph" in *The Scribner Encyclopedia of American Lives: The 1960s*, ed. William L. O'Neill and Kenneth T. Jackson (New York: Charles Scribner's, 2003), 607–608; Wollman and Inman, *Portraits in Steel*, 172–185; Hoerr, *And the Wolf Finally Came*, 418–419; "J&L to Recall," *PPG* 12/30/71.

10. On November 11, 1971, the board denied the Clairton slag quenching variance, but U.S. Steel quickly appealed the decision to the board, even though that was not an option for appeal in Article XVII; the matter was thus unresolved.

11. "Variance Compliance, 1970–1971," ff "Variance Board," GASP Internal; Longhurst, "Don't Hold Your Breath," table 5.2.

12. Telegram, Madoff to Dunsmore, October 21, 1971, ff "Correspondence," GASP Internal. For Dunsmore's transition from county regulator to employee of U.S. Steel, see "Smoke Control Chief," *PPG* 6/3/65; "Pollution Number Game," *PP* 11/21/69.

13. "Pollution Appeal Hit," *PPG* 2/19/71; "Clean Air Panel Raps Coke Plant," *PPG* 11/15/71; "U.S.S. Clairton Works Spew," *PF* 10/29/71; "U.S. Steel Accused by Pollution Board," *NYT* 11/15/71.

14. See docket no. 143, "Variance Compliance, 1970–1971," ff "Variance Board," GASP In-

ternal. One meeting to plan strategy took place on December 20, 1971, as described in GASP *Hotline* 2 (March–April 1972): 3

15. Warren, *Big Steel*, 307.

16. *County of Allegheny vs. United States Steel*, Allegheny Court of Common Pleas, Civil Division, docket no. 1550, April Term 1972, 12 and 20–22. Materials related to this case are held by the Court of Common Pleas, and are hereinafter referred to as *County vs. U.S. Steel.*

17. "Press Conference . . .," GASP *Hotline* (March–April 1972): 4; Silvestri joined the court in 1968. Pamphlet, "The Court of Common Pleas of Allegheny County, Pennsylvania," prepared by the Academy of Trial Lawyers of Allegheny County, May 1973, in the University of Pittsburgh's Balco Law Library, State Materials Collection; Joel Fishman, *Judges of Allegheny County, Fifth Judicial District, Pennsylvania: 1788–1988* (Pittsburgh: Allegheny County Court of Common Pleas Bicentennial Commission, 1988), 144.

18. Fishman, *Judges of Allegheny County.*

19. "'Dirty Movie' Features Pollution," *PP* 2/16/72; "Statement of Mrs. John B. Newman, February 15, 1972," press release, "Address to the State Legislature of Pennsylvania," both from ff "Madoff & Newman," GASP Internal; Lave quote from "' Dirty Movie.'"

20. "Dirtie Gertie—Be My Valentine" GASP *Hotline* (March–April 1972): 5; "Landmark Suit against Coke Plant Expected by County Bureau, GASP," *MDN* n.d., likely April, 1972. Copy in author's possession.

21. Madoff to Earl W. Mallick, April 17, 1972, ff "Task Force Report," GASP Internal.

22. "GASP Task Force Report, Clairton Coke Works, U.S. Steel" (June 1972), ff "Task Force Report," GASP Internal, hereinafter "Task Force Report"; GASP Board of Directors' Minutes, June 1972, 3, CLP GASP.

23. "Report III: Treatment of Coke Waste Liquors," "Task Force Report." Ed Swanson's resume is in ff "GASP Staff Bios," GASP Internal. "Introduction and Perspective," "Task Force Report"; ff "USS/Purloined Letter," GASP Internal.

24. Madoff interview with author, Surprise, Arizona, October 15–16, 2007.

25. Silvestri to Madoff, November 20, 1972; Demase to Cardinal, August 1, 1972; Fein to GASP Board of Directors, July 11, 1972, all in ff "Task Force Report," GASP Internal; "GASP vs. U.S. Steel," *PF* 6/30/72.

26. For increasing pressure from labor unions, U.S. Steel's Gary works, and a DER suit over U.S. Steel Homestead, see "Coke Works Cited," *PP* 6/22/72; "Court Orders Air Controls," *PPG* 5/24/72. See also Andrew Hurley, *Environmental Inequalities: Class, Race and Industrial Pollution in Gary, Indiana, 1945–1980* (Chapel Hill: University of North Carolina Press, 1995), 158–162; *United States Steel Corporation v. Department of Environmental Resources, Pennsylvania Commonwealth Court Reports* 7 (Sayre, PA: Murrelle, 1973), 429.

27. "U.S. Steel Bows," *NYT* 8/21/72; cf. "U.S. Steel Forced," *NYT* 8/27/72.

28. Consent decree (1972), *County vs. U.S. Steel.* The requirement for EPA approval seemed to be a formality for the local press: "Supposedly, arrangements have been worked out that will enable EPA to approve the pact or to refrain from taking legal action against U.S. Steel so long as it abides by the terms of the decree." *PP* 9/25/72; quoted in Jones, *Clean Air*, 282.

29. Consent decree, § 4.

30. Editorial, "Significance of Clairton Cleanup," *PPG* 10/27/72.

31. "U.S. Steel and Environment," *PF* 9/1/72; "U.S. Steel Forced," *NYT* 8/27/72; see also "U.S. Steel Agrees," *WSJ* 9/26/72; "Steel's Big Cleanup," *Newsweek* (September 4, 1972).

32. "Hunt Scores U.S. Rejection," *PP* 10/29/72; Jones, *Clean Air*, 284–285

33. "Hunt Scores"; Jones, *Clean Air*, 283–284.

34. GASP *Hotline* (November 1972): 6; cf. Jones, *Clean Air*, 286; Madoff interview with author, Surprise, Arizona, October 15–16, 2007.

35. "Significance of Clairton," *PPG* 10/27/72, 10; "Genesis II," *PPG* 3/28/73; cf. *NYT* coverage *supra*; "Steel's Big Cleanup," *Newsweek* (September 4, 1972), 62; Jones, *Clean Air*, 285; "Calming the Clairton Fuss," *PPG* 10/9/72.

36. "When Oh When Oh U.S. Steel," GASP *Hotline* (October 1973): 6; Jones, *Clean Air*, 287; "After Gagging USS Clairton Critic," *PP* 6/14/73.

37. "Pollution Levels," *PPG* 1/20/73; Jones, *Clean Air*, 287–288; GASP *Citizen's Commentary* (July 1973): 3, GASP Internal.

38. Jones, *Clean Air*, 289; *PP* 1/18/73; Bernard Bloom to Gerald P. Dodson, February 20, 1973, ff "USS/Clairton 1973," GASP Internal. Emphasis in original. For the many disputes at Clairton, see among others "Airing Ruled for Clairton Smokestacks," *PPG* 11/8/73; "USS Gets Word," *PP* 4/19/73; Council on Economic Priorities press release, "Steel Industry Fails to Abate Pollution," May 22, 1973, and GASP press release, May 22, 1973, ff "USS/Clairton 1973," GASP Internal; "Area Steel Mills Cited," *PPG* 5/22/73.

39. *Commonwealth of Pennsylvania, Appellant, v. United States Steel Corporation, Appellee, County of Allegheny, Appellant, v. United States Steel Corporation, Appellee*, nos. 960 and 961 C. D. 1973, hereinafter referred to as *Commonwealth and County v. U.S. Steel*.

40. "Clairton Cleanup Tab," *PP* 4/13/73; quote from "Fine USS $300,000, State Asks," *PP* 4/17/73.

41. *Commonwealth and County v. U.S. Steel*; "USS Cleared," *PP* 5/23/73.

42. "In Silvestri Court," *PP* 6/10/73; Gorr to Staisey, Foerster, and Hunt, June 7, 1973, ff "USS/Clairton 1973," GASP Internal; "County Appeals USS Air Ruling," *PP* 6/11/73.

43. "Coke Battles on New Kick," *PP* 6/16/73; "Judge Still Refuses," *PP* 6/26/73.

44. Gorr to Robert W. Fri, acting director, EPA, June 25, 1973, ff "USS/Clairton 1973," GASP Internal.

45. Gorr to Chleboski, March 29, 1973; Gorr to Bloom and Dodson, March 28, 1973, both in ff "USS/Clairton 1973," GASP Internal.

46. Mary Brignano and J. Tomlinson Fort, *Reed Smith: A Law Firm Celebrates 125 Years* (Pittsburgh: Reed Smith LLP, 2002).

47. GASP *Hotline* 3 (May 1973); *Commonwealth and County v. U.S. Steel*; "Appeals Next Step," *PP* 6/12/73; "After Gagging USS," *PP* 6/14/73.

48. "State Says . . . ," GASP *Hotline* (October 1973): 7.

49. Daniel J. Snyder III replaced Furia as Region III director in May 1973. "EPA Region III Director Profile," GASP *Hotline* (January 1974): 9; "When Oh When Oh U.S. Steel," GASP *Hotline* (July–August 1974): 7; "E.P.A. Lists 63 Clean-Air Act Violations," *NYT* 11/9/73.

50. "Citizens Demand USSteel Cleanup," *PPG* 11/10/73; "GASP Says U.S. Steel Trying to Buy Time," *MDN* 11/10/73; "State Rejects USS," *PP* 11/28/73.

51. Pennsylvania House Bills 1425-1427; "Open Door to Pollution," *PP* 2/13/74; "Clear Environmental Threat," *PP* 3/10/74; "Bethlehem Bills Draw Fire," *PP* 3/15/74; "Scrap the Bethlehem Bills," *PPG* 3/19/74; "Schapp Backs Mills Plan," *PPG* 3/29/74.

52. "U.S. Steel's New Commitment, *PPG* 6/12/74; "2 U.S. Steel," *PP* 3/29/74; "USS Gets Delay," *PP* 9/30/74; "Scrap the Bethlehem Bills"; "Bethlehem Bills Draw Fire."

53. GASP *Hotline* 4 (July–August 1974): 1.

54. The Commonwealth Court decision in *Commonwealth and County v. U.S. Steel* can be found at 15 Pa. Commw. 184, 325 A. 2d 324, 1974 Pa. Commw. Lexis 707, decided September 6, 1974. U.S. Steel appealed the decision to the state Supreme Court, which refused to hear the case. "U.S. Steel under Fire," *NYT* 7/13/75.

55. Gorr to Chleboski, September 11, 1974, ff "Gorr Correspondence," GASP Internal; *In Re Grand Jury Proceedings*, 75-1450 and 75-1456, U.S. Third Circuit Court of Appeals, October 20, 1975.

56. GASP *Hotline* (November 1974).

57. Opinion, June 4, 1975, *County v. U.S. Steel.*

58. *NYT* 7/13/75.

59. "County's Silvestri Critic," *PP* 9/17/75.

60. "Ellenbogen to Settle USS," *PP* 9/17/75; "State Charges Called Smear," *PPG* 10/22/75. Ellenbogen's background is in Joel Fishman, *Judges of Allegheny County*, 144; quote from Ellenbogen Opinion, October 24, 1975, *County v. U.S. Steel.*

61. Madoff interview with author, Surprise, Arizona, October 15–16, 2007.

62. "County's Silvestri Critic."

63. Quote from Ellenbogen decision, December 30, 1976, *County vs. U.S. Steel*; "U.S. Steel Signs," *NYT* 10/12/76; "U.S. Steel Agrees," *WSJ* 10/12/76; "USS Plans $600 Million Cleanup," *PP* 4/11/76; ff "USS/Clairton 1973," GASP Internal; "Fearful of Being Ignored," *PPG* 10/2/76.

64. "Non-Enforcement and the Double Standard," GASP *Hotline* (July 1973): 3.

65. GASP *Hotline* (July–August 1974): 1; "J&L May Abandon," *PP* 10/17/74; "Air Laws Blamed," *PP* 11/15/74; "Loss of J&L Jobs," *PPG* 1/15/75.

66. "GASP Sues J&L on Pollution Control," *PP* 3/24/75; GASP press release, Monday, March 24, 1975, ff 16, AIS ACAPCAC; "Clean Air Timetable Signed," *PP* 10/8/75.

67. "Jobs Loss Feared, Air Variance Given," *PPG* 10/7/70.

68. Letters to the editor, *PPG* 4/3/75.

69. "Clean Air Panel Raps Coke Plant," *PPG* 11/15/71; "USSteel Hit by Silvestri," *PPG* 5/24/73; "Rigid Air Pollution Laws," *PP* 4/17/73.

70. Film, *Don't Hold Your Breath, Fight for It*, in AIS Pelkofer; Gorr to John DeFazio, November 7, 1974, "ff Gorr Correspondence," GASP Internal. See also letter to the editor, *PP*, 1/10/75; letter to the editor, *PPG* 4/24/75; "J&L Case of Possible Jobs v. Pollution," *PP* 1/8/75.

71. Manuscript, n.d., likely October 1974, ff "Gorr Correspondence," GASP Internal. Gorr cites "Scapegoat for Inflation," *NYT* 9/23/74.

72. Advertisement, *PPG* 5/18/76; "A Steely Approach," *PP* 5/19/76; "Enviro-SOS Ad Counters," *PP* 6/29/76; Advertisement, *PP* 6/29/76.

73. Warren, *Big Steel*, 307; cf. discussion of Speer's comments on jobs in Pittsburgh in "Clairton: Clean or Close?" *Iron Age* (October 25, 1976): 9; James Vallela, "May Quit Mon Valley," *PP*, n.d., ff "USS/Clairton 1973," GASP Internal; "Speer Tough on Imports, Pollution," *PP* 5/3/77. Cf. the 1977 debate between the EPA and U.S. Steel over the Homestead Works, "Pierce Vows War," *PP* 8/28/77; "Closing Furnace Here," *PP* 7/31/77. On Speer's language, see, among many, "U.S. Steel Disputed," *PP* 5/22/77; letter to the editor, "U.S. Steel Claim 'Patently False,'" *PPG* 5/25/77. GASP president Walter Zadan wrote to DER general counsel William M. Eichbaum to thank Eichbaum for responding "to the latest blackmail attempt from Edgar Speer" in a letter to the editor, *PPG* 5/25/77. Letter, Zadan to Eichbaum, n.d., ff "USS/Clairton 1973," GASP Internal; City of Clairton Council Resolution No. 840, April 24, 1979, ff "USS/Clairton 1," GASP Internal.

74. For COKE, see statement of Priscilla McFadden, acting chairman of COKE, November 22, 1976, ff "USS/Clairton 1973," GASP Internal; "USW Aide," *PP* 11/23/76; see also "300 Attend Hearing," *Pennsylvania EcoNotes* (November 1976): 5; "Coking Decree Hearing," *MDN* 11/23/76.

75. Quotes from letter to the editor, *PPG*, 11/6/75 and 12/3/75; cf. Gorr's definitive response 11/15/75.

76. "Citizens Wise Up," *PP* 11/26/72; "Anti-Pollution Drive," *PP* 3/13/72.

77. GASP did continue to reference children's health, but most discussion of U.S. Steel stuck to the themes of law enforcement and citizen's rights. See "School Children Affected," *Hotline* (July 1973): 6; GASP film, *Don't Hold Your Breath, Fight for it.*

78. GASP press release, "GASP Applaud's EPA's Efforts to Bring Clairton Works Into Compliance," November 9, 1973, ff "USS/Clairton 1973," GASP Internal; *PP* 6/14/73; McHugh quote from "A Steely Approach," *PP* 5/19/76; Cardinal to Staisey, Foerster, and Hunt, April 18, 1973, ff "USS/Clairton 1973," GASP Internal.

79. "State Rejects USS Clairton Plant Air Plan," *PP* 11/28/73.

80. Jeannette Widom, "Party Cookies Only," GASP pamphlet, 1972, vi–vii.

81. "U.S. Steel Signs Accord," *NYT* 10/12/76; "U.S. Steel Agrees," *WSJ* 10/12/76; "Steel's Trade-Off on Coke Emissions," *Business Week* (November 1, 1976): 26; "U.S. Steel Signs Accord," *WSJ* 12/31/76, 3; *United States of America, et al., v. United States Steel Corporation,* Civil Action 79-709, U.S. District Court for the Western District of Pennsylvania, 1983 U.S. dist Lexis 13841, 16 ERC (BNA) 1228, 15 ELR 21012; *United States of America, et al., v. United States Steel Corporation,* Civil Action 79-709, United States District Court for the Western District of Pennsylvania, 87 F.R.D. 709, 1980 U.S. Dist Lexis 17348, 30 Fed. R. Serv. 2d (Callaghan) 742, 16 ERC (BNA) 1228. For a timeline of plant closures in the Pittsburgh region, see Hoerr, *And the Wolf Finally Came,* 10-11. An analysis of the comparative impacts of air pollution regulation and economic change is available from Mark R. Powell, "Three-City Air Study," Resources for the Future Discussion Paper 97-29 (Washington, DC: Resources for the Future, 1997), 15, available at http://www.rff.org/

82. "Out of Funds," *PP* 3/25/81; *United States of America, v.* LTV Steel Company, Inc., Civil Action 98-570, 187 F.R.D. 522, E.D. Pa., 1998. See *United States of America, et al., v. LTV Steel Company, Inc,* Civil Action 98-570, 116 F. Supp 2d 624, 2000 U.S. Dist Lexis 14660, 51 ERC (BNA) 1496, 31 ELR 20232, September 29, 2000; and that same case, 269 B.R. 576, 2001 U.S. Dist. Lexis 18367, November 7, 2001.

83. *The Group Against Smog and Pollution, Inc., et al., v. United States Environmental Protection Agency,* No. 78-1534, U.S. Court of Appeals, District of Columbia Circuit, 214 U.S. App. DC 466, 665 F. 2d 1284; 1981 U.S. App Lexis 17458; 16 ERC (BNA) 1917; 11 ELR 20982.

84. *Sierra Club v. Whitman,* D.D.C., 2002, March 11, 2002, 32 *Environmental Law Reporter* 20, 538.

85. Cary Coglianese, "Litigating within Relationships: Disputes and Disturbance in the Regulatory Process," *Law and Society Review* 30 (1996): 765.

86. "U.S. Steel Agrees," *PPG* 06/08/07; "DEP May Get Control," *PTR* 11/08/07; "U.S. Steel Puts Hold," *PPG* 04/02/09; "Clairton, Glassport Cancer Risk," *PPG* 6/24/09; editorial, *PPG* 6/28/09.

Conclusion

1. For a summary of these approaches, and the importance of contemporaneous events, see Adam W. Rome, "'Give Earth A Chance': The Environmental Movement and the Sixties," *Journal of American History* 90 (2003): 526 n. 3.

2. For the argument that modern environmentalism is a continuation of older concerns, see Robert Gottlieb's many works, starting with *Forcing the Spring: The Transformation of the American Environmental Movement* (Washington, DC: Island Press, 1993).

3. George Hoberg, *Pluralism by Design: Environmental Policy and the American Regulatory State* (New York: Praeger, 1992), chapters 2 and 3; Richard N. L. Andrews, *Managing the Environment, Managing Ourselves: A History of American Environmental Policy* (New

Haven: Yale University Press, 1999), 360; Richard Munton, "Deliberative Democracy and Environmental Decision-Making," in *Negotiating Environmental Change: New Perspectives from Social Science*, ed. Frans Berkhout, Melissa Leach, and Ian Scoones (Cheltenham, UK: Edward Elgar, 2003), 109; National Environmental Policy Act of 1969, at 42 U.S.C. 4321–4361; Serge Taylor, *Making Bureaucracies Think: The Environmental Impact Statement Strategy of Administrative Reform* (Stanford, CA: Stanford University Press, 1984), 7, 295; Frank Uekoetter, *The Age of Smoke: Environmental Policy in Germany and the United States, 1880–1970* (Pittsburgh: University of Pittsburgh Press, 2009), 238–246; Noga Morag-Levine, "Partners No More: Relational Transformation and the Turn to Litigation in Two Conservationist Organizations," *Law & Society Review* 37 (June 2003): 457–510.

4. *I Belong Here!* GASP (1972), sixteen mm film, CLP GASP.

5. "County Air Shows Vast Improvement," *PPG* 9/14/78.

6. "The Greening of Pittsburgh," *NYT* 3/31/09.

7. "Clairton, Glassport Cancer Risk," *PPG* 6/24/09; "Risky Business," editorial, *PPG*, 6/26/09; "Region Passes L.A.," *PTR* 5/1/08.

8. Michael Shellenberger and Ted Nordhaus, "The Death of Environmentalism: Global Warming Politics in a Post-Environmental World," unpublished manuscript, 2004, available from http://www.thebreakthrough.org/; see also "Paper Sets Off a Debate," *NYT*, 2/6/05; Nicholas Kristof op-ed, "'I Have a Nightmare,'" *NYT*, 3/12/05. See also Samuel P. Hays, "Beyond Celebration: Pittsburgh and Its Region in the Environmental Era—Notes by a Participant Observer," in *Devastation and Renewal: An Environmental History of Pittsburgh and Its Region*, ed. Joel A. Tarr (Pittsburgh: University of Pittsburgh Press, 2005), 193–215.

9. "Novel Antipollution Tool," *NYT* 6/4/99; Sharon Beder, *Global Spin: The Corporate Assault on Environmentalism* (White River Junction, VT: Chelsea Green, 2002), chapter 4; David P. Gionfriddo, "Sealing Pandora's Box: Judicial Doctrines Restricting Public Trust Citizen Environmental Suits," *Boston College Environmental Affairs Law Review* 13 (Spring 1986): 439; Sam Kalen, "Standing on Its Last Legs: *Bennett V. Spear* and the Past and Future of Standing in Environmental Cases," *Journal of Land Use and Environmental Law* 13 (Fall 1997): 1; Gene R. Nichol, "Citizen Suits and the Future of Standing in the 21st Century: From Lujan to Laidlaw and Beyond: The Impossibility of Lujan's Project," *Duke Environmental Law and Policy Forum* 11 (Spring 2001): 193. The two most important Supreme Court decisions are *Steel Co. v. Citizens for a Better Environment*, 118 S. Ct. 1003 (1998), and *Lujan v. Defenders of Wildlife*, 504 U.S. 555 (1992).

10. Cornelia Dean, "Report Says Public Outreach, Done Right, Aids Policymaking," *NYT*, 8/22/08; Thomas C. Beierle and Jerry Cayford, *Democracy in Practice: Public Participation in Environmental Decisions* (Washington, DC: Resources for the Future, 2002), 74–76; W. Michelle Simmons, *Participation and Power: Civic Discourse in Environmental Policy Decisions* (Albany: SUNY Press, 2007), 1–23; EPA Office of Inspector General Audit Report, "Public Participation in Louisiana's Air Permitting Program and EPA Oversight," Report No. 01351-2002-P-00011, August 7, 2002; Frans H. J. M. Coenen, Dave Huitema, and Laurence J. O'Toole, eds., *Participation and the Quality of Environmental Decision Making* (Dordrecht: Kluwer Academic, 2000). For an opposing argument, see George A. Gonzalez, *The Politics of Air Pollution: Urban Growth, Ecological Modernization, and Symbolic Inclusion* (Albany: SUNY Press, 2005), 14–16, 24–26, 89–107.

11. Tom Imerito, "The Politics of Smoke," *Pittsburgh Pulp* (February 26, 2004); GASP *Hotline* (Winter 2009–Spring 2009).

12. "GASP Only Public Group to Show at Smog Hearing," *PPG* 6/4/81; "GASP Endorses," *PPG* 2/21/79.

13. "DEP May Get Control," *PTR* 11/8/07; "County to Keep Air Enforcement," *PTR* 4/12/08; "Allegheny County Board of Health Revises Changes," *PTR* 5/7/09; "Last Gasp for Allegheny County's," *PPG* 11/4/07; editorial, "Quality Air," *PPG* 11/9/07; "Onorato: Pollution Control Chief Wasn't Forced Out," *PPG* 2/12/08.

Bibliography

A Note on Archival Sources

It is both a blessing and a curse to write about comparatively recent historical events. The blessing is the richness and variety of the available documents. The curse is that these records have not yet been fully indexed, organized, surveyed, or in some cases even placed in appropriate archives.

Three archival sources used in this manuscript require some explanation. First, I indexed the large collection of GASP correspondence, minutes, and publications donated to the Carnegie Library of Pittsburgh during an early stage of this project. The library subsequently donated those records to the Archives Service Center at the University of Pittsburgh, where they will eventually be added to the GASP materials already there. Second, the records cited in this work as "GASP Internal" were made available by GASP, are currently in my possession, and will be donated to the Archives Service Center at the conclusion of this project. Finally, many of the records held by the National Archives and Records Administration under Record Group 412 have not yet been fully indexed, organized, or placed in appropriate archival containers. In the case of these three sources, location information cited in this work is likely to change as the records continue to be processed.

Where appropriate, the list of archival sources below begins with the abbreviation used in notes to indicate the name of the source. In the notes themselves, ff indicates "file folder." Some archival collections or manuscript series listed below provided background or supporting material; some were consulted to verify that they did not contain contradictory information. Those collections are listed here but are not specifically cited in individual notes.

Archival Sources

ALLEGHENY COUNTY COURT OF COMMON PLEAS, DEPARTMENT OF COURT RECORDS, CITY-COUNTY BUILDING, 414 GRANT STREET, PITTSBURGH, PA

County vs. USS: Papers related to docket 1550, April term 1972, Civil Division, *County of Allegheny vs. United States Steel.*

ALLEGHENY COUNTY HEALTH DEPARTMENT, AIR QUALITY PROGRAM LIBRARY, 301 39TH STREET, BLDG. #7 PITTSBURGH, PA

ACHD BAPC: Records of the Allegheny County Health Department Bureau of Air Pollution Control.

ARCHIVES OF INDUSTRIAL SOCIETY, ARCHIVES SERVICE CENTER, UNIVERSITY OF PITTSBURGH, 7500 THOMAS BOULEVARD, PITTSBURGH, PA

AIS ACAPCAC: Allegheny County Air Pollution Control Advisory Committee, 92:3.

AIS ACHD BAPC: Allegheny County Health Department, Bureau of Air Pollution Control, Material on Smoke Control and Air Pollution, 1941–1978, 80:7.

AIS Broughton: Robert F. Broughton Papers, 77:42.

AIS GASP: Group Against Smog and Pollution, Air Pollution Variance Petition Files, 1972–1975, 79:21.

AIS Hays: Samuel P. Hays Papers, 91:9.

AIS J&L: Jones & Laughlin Steel Corporation, Pittsburgh, PA, Records, ca 1900–1970, 88:7.

AIS Madoff: Michelle Madoff Papers, 1970–1977, 78:3.

AIS Pelkofer: Patricia Pelkofer Papers, 93:5.

AIS Schlesinger: Hymen Schlesinger Papers, 1944–1975, 77:33.

AIS Widom: Jeannette Widom Papers, 1970–1981, 94:10.

CARNEGIE LIBRARY OF PITTSBURGH, PENNSYLVANIA DEPARTMENT, PITTSBURGH, PA

CLP GASP: Papers of the Group Against Smog and Pollution; now in AIS.

GROUP AGAINST SMOG AND POLLUTION, GASP OFFICES, WIGHTMAN SCHOOL COMMUNITY BUILDING, 5604 SOLWAY STREET, #204, PITTSBURGH, PA

GASP Internal: Records of the Group Against Smog and Pollution; currently in author's possession.

HISTORICAL SOCIETY OF WESTERN PENNSYLVANIA, PITTSBURGH, PA

HSWP GASP: Group Against Smog and Pollution (GASP), Records, 1969–1988, MSS #43.

HSWP Venable: Emerson Venable Papers, MSS #347.

NATIONAL ARCHIVES AND RECORDS ADMINISTRATION, COLLEGE PARK, MD

NARA RG 412: Records of the Environmental Protection Agency and Predecessor Organizations, Record Group 412.

PENNSYLVANIA STATE ARCHIVES, HARRISBURG, PA

PSA RG 11.41: Record Group 11.41, Office of Legal Counsel.

PSA RG 11.42: Record Group 11.42, Air Pollution Commission.

PSA RG 11.43: Record Group 11.43, Air Pollution Commission Transcripts of Hearings and Position Papers on Proposed Regulations.

PSA RG 43.16: Record Group 43.16, Bureau of Air Quality Control—Hearing Transcripts and Position Papers of the Environmental Quality Board, 1971–1972.

PSA RG 43.35: Record Group 43.35, Environmental Hearing Board Adjudications, 1971–1984.

PSA RG 43.36: Record Group 43.36, Environmental Quality Board Minutes, 1971–1983.

PSA RG 43.37: Record Group 43.37, Environmental Quality Board Public Hearing Transcripts, 1971–1982.

Newspapers

Newspaper accounts are integral to this work, and I have accumulated more than five hundred individual stories, mostly from the *Pittsburgh Press* and *Pittsburgh Post-Gazette* in the 1960s and 1970s. Not all of these newspapers have been cited in the text, even if the name of the paper appears in this list. No index for these local papers yet exists for this time period, so the stories were collected from clippings files in archives or in the possession of GASP members and from microfilm in the Carnegie Library of Pittsburgh and the State Library in

Harrisburg. With the goal of making the notes section more compact, full citations of articles from frequently cited newspapers have been summarized to a shortened headline, an abbreviated newspaper name, and month/date/year where possible. Any abbreviations used appear below.

Allegheny County *Advance-Leader*
Donora Herald-American, 1940–1970; later *Monongahela Daily Herald*, 1970–
Homestead Messenger, 1968–1972
MDN: McKeesport Daily News, 1970–1977
Monroeville News
NYT: New York Times, 1960–1999
North Hills News Record
WSJ: Wall Street Journal, 1960–1979
Pittsburgh Catholic
Pittsburgh City Paper, 2003–2009
Pittsburgh Courier (also known as *The New Pittsburgh Courier*), 1965–1975
PF: Pittsburgh Forum, 1971–1976
PP: Pittsburgh Press, 1959–1989
PPG: Pittsburgh Post-Gazette, 1959–2003
Pittsburgh *Pulp*, 2002–2004
Pittsburgh *Sun-Telegraph*
PTR: Pittsburgh Tribune-Review, 2007–2009
Sewickley *Herald*
Squirrel Hill *News*

Select Bibliography

This is a short list of useful secondary sources. This bibliography is not comprehensive, and mention of an author's seminal work should be read as standing for the author's wider catalog. Most works contemporaneous with the events in the text are found in the notes and not in this bibliography. I have roughly divided this list into Air Pollution; Civil Society and Cities; Environmental History; Pittsburgh and Pennsylvania; and U.S. History, Women and the Long 1960s.

Air Pollution

Aiken, Katherine G. "'Not Long Ago a Smoking Chimney Was a Sign of Prosperity': Corporate and Community Response to Pollution at the Bunker Hill Smelter in Kellogg, Idaho." *Environmental History Review* 18 (Spring 1994): 67–86.

Bernstein, Shawn. "The Rise of Air Pollution Control as a National Political Issue." Ph.D. dissertation, Columbia University, 1986.

Berry, Michael Arnold. "A Method for Examining Policy Implementation: A Study of Decisionmaking for the National Ambient Air Quality Standards, 1964–1984." Ph.D. dissertation, University of North Carolina, Chapel Hill, 1984.

Brown, Lynne Anne Baker. "The Clean Air Amendments of 1977: Public Policy Implementation at the Local Level of Government." Master's thesis, University of Pittsburgh, 1981.

Bryson, Chris. "A Secret History of America's Worst Air Pollution Disaster." *Earth Island Journal* 13 (Fall 1998): 36.

Crenson, Mathew A. *The Un-Politics of Air Pollution: A Study of Non-Decisionmaking in the Cities.* Baltimore: Johns Hopkins University Press, 1971.

Davidson, Cliff. "Air Pollution in Pittsburgh: Historical Perspective." *Journal of the Air Pollution Control Association* 29, 10 (1979): 1035–1041.

Davis, Devra. *When Smoke Ran Like Water: Tales of Environmental Deception and the Battle against Pollution.* New York: Basic Books, 2002.

Dewey, Scott H. *Don't Breathe the Air: Air Pollution and U.S. Environmental Politics, 1945–1970.* College Station: Texas A&M University Press, 2000.

Dingle, A. E. "'The Monster Nuisance of All': Landowners, Alkali Manufacturers, and Air Pollution, 1828–64." *Economic History Review* 35 (November 1982): 529–548.

Gonzalez, George A. *The Politics of Air Pollution: Urban Growth, Ecological Modernization, and Symbolic Inclusion.* Albany: SUNY Press, 2005.

Grinder, Robert Dale. "The Anti Smoke Crusades: Early Attempts to Reform the Urban Environment, 1893–1918." Ph.D. dissertation, University of Missouri-Columbia, 1973.

Gugliotta, Angela. "'Hell with the Lid Taken Off': A Cultural History of Air Pollution—Pittsburgh." Ph.D. dissertation, University of Notre Dame, 2004.

Haskell, Elizabeth. *The Politics of Clean Air: EPA Standards for Coal-Burning Power Plants.* Westport, CT: Praeger, 1982.

Jacobs, Chip, and William J. Kelly. *Smogtown: The Lung-Burning History of Pollution in Los Angeles.* Woodstock, NY: Overlook Press, 2008.

Jacobson, Mark Zachary. *Atmospheric Pollution: History, Science, and Regulation.* New York: Cambridge University Press, 2002.

Jenner, Mark S. R. "Underground, Overground: Pollution and Place in Urban History." *Journal of Urban History* 24, 1 (1997): 97–110.

Johnson, Janet Buttolph. "An Airing of the Clean Air Act." *Journal of Politics* 47 (February 1985): 292–304.

Jones, Charles O. *Clean Air: The Policies and Politics of Pollution Control.* Pittsburgh: University of Pittsburgh Press, 1975.

Lave, Lester B., and Gilbert S. Omenn. *Clearing the Air: Reforming the Clean Air Act.* Washington, DC: Brookings Institution, 1981.

Meetham, A. R. *Atmospheric Pollution: Its History, Origins and Prevention.* Oxford: Pergamon Press, 1981.

Melichar, Kenneth Edward. "The Making of the 1967 Montana Clean Air Act: A Struggle over the Ownership of Definitions of Air Pollution." *Sociological Perspectives* 30, 1 (1987): 49–70.

Melosi, Martin V. *Effluent America: Cities, Industry, Energy, and the Environment.* Pittsburgh: University of Pittsburgh Press, 2001.

Morag-Levine, Noga. *Chasing the Wind: Regulating Air Pollution in the Common Law State.* Princeton: Princeton University Press, 2003.

Pearson, James. *Reducing the Averages: The Founding and Development of the Puget Sound Air Pollution Control Agency.* San Jose, CA: Writer's Club Press, 1999.

Pittman, Walter E., Jr. "The Smoke Abatement Campaign in Salt Lake City, 1890–1925." *Locus* 2 (Fall 1989): 69–78.

Platt, Harold L. "Invisible Gases: Smoke, Gender, and the Redefinition of Environmental Policy in Chicago, 1900–1920." *Planning Perspectives* 10 (1995): 67–97.

Reitze, Arnold W., Jr. *Air Pollution Control Law: Compliance and Enforcement.* Washington, DC: Environmental Law Institute, 2001.

Ringquist, Evan J. "Is 'Effective Regulation' Always Oxymoronic? The States and Ambient Air Quality." *Social Science Quarterly* 76, 1 (1995): 69–87.

Rosen, Christine Meisner. "Businessmen against Pollution in Late Nineteenth Century Chicago." *Business History Review* 69 (Autumn 1995): 351–397.

Schulze, R. H. "The 20-Year History of the Evolution of Air Pollution Control Legislation in the U.S.A." *Atmospheric Environment* 27B (March 1993): 15–22.

Seinfeld, John H. "The Dearth of the Clinic: Lead, Air, and Agency in Twentieth-Century America." *Journal of the History of Medicine* 58 (July 2003): 255–291.

Snyder, Lynne Page. " 'The Death-Dealing Smog over Donora, Pennsylvania': Industrial Air Pollution, Public Health Policy, and the Politics of Expertise, 1948–1949." *Environmental History Review* 18 (Spring 1994): 117.

Stradling, David. *Smokestacks and Progressives: Environmentalists, Engineers and Air Quality in America, 1881–1951.* Baltimore: Johns Hopkins University Press, 1999.

Uekoetter, Frank. *The Age of Smoke: Environmental Policy in Germany and the United States, 1880–1970.* Pittsburgh: University of Pittsburgh Press, 2009.

Wark, Kenneth, Cecil F. Warner, and Wayne T. Davis, eds. *Air Pollution: Its Origins and Control.* 3rd ed. Menlo Park, CA: Addison Wesley Longman, 1998.

Wood, B. Dan. "Modeling Federal Implementation as a System: The Clean Air Case." *American Journal of Political Science* 36 (February 1992): 40–67.

Civil Society and Cities

Ahlbrandt, Roger S. Jr. *Neighborhoods, People, and Community.* New York: Plenum, 1984.

Almond, Gabriel A., and Sidney Verba. *The Civic Culture: Political Attitudes and Democracy in Five Nations.* Boston, MA: Little, Brown, 1963.

Banfield, Edward C., and James Q. Wilson. *City Politics.* Cambridge, MA: Harvard University Press, 1967.

Brehm, John, and Wendy Rahn. "Individual-Level Evidence for the Causes and Consequences of Social Capital." *American Journal of Political Science* 41 (July 1997): 999–1023.

Calhoun, Craig. "Civil Society and the Public Sphere." *Public Culture* 5 (1993): 267–280.

Charney, Evan. "Political Liberalism, Deliberative Democracy, and the Public Sphere." *American Political Science Review* 92 (March 1998): 97–110.

Clemens, Elizabeth S. *The People's Lobby: Organizational Innovation and the Rise of Interest Group Politics in the United States, 1890–1925.* Chicago: University of Chicago Press, 1997.

Cole, Richard L. *Citizen Participation and the Urban Policy Process.* Lexington, MA: Lexington Books, 1974.

Coleman, James S. *Foundations of Social Theory.* Cambridge, MA: Harvard University Press, 1990.

Dahl, Robert A. *Who Governs? Democracy and Power in an American City.* New Haven: Yale University Press, 1961.

Ethington, Philip J. *The Public City: The Political Construction of Urban Life in San Francisco, 1850–1900.* New York: Cambridge University Press, 1994.

Fairbanks, Robert B., and Patricia Mooney-Melvin, eds. *Making Sense of the City: Local Government, Civic Culture, and Community Life in Urban America.* Columbus: Ohio State University Press, 2001.

Foley, Michael W., and Bob Edwards. "The Paradox of Civil Society." *Journal of Democracy* 7, 3 (1996): 38–52.

Galston, William A. "Civil Society and the 'Art of Association.' " *Journal of Democracy* 11, 1 (2000): 64–70.

Habermas, Jurgen. *The Structural Transformation of the Public Sphere: An Inquiry into a Category of Bourgeois Society.* Cambridge, MA: Massachusetts Institute of Technology Press, 1996.

Johnson, James, and Dana R. Villa. "Public Sphere, Postmodernism and Polemic." *American Political Science Review* 88 (June 1994): 427–433.

Judge, David, Gerry Stoker, and Harold Wolman, eds. *Theories of Urban Politics*. Thousand Oaks, CA: Sage, 1995.

Kornbluh, Mark Lawrence. *Why America Stopped Voting: The Decline of Participatory Democracy and the Emergence of Modern American Politics*. New York: New York University Press, 2000.

Marston, Sallie A. "Citizen Action Programs and Participatory Politics in Tucson." In *Public Policy for Democracy*, edited by Helen Ingram and Steven Rathgeb Smith, 119–135. Washington, DC: Brookings Institution, 1993.

Mattson, Kevin. *Creating a Democratic Public: The Struggle for Urban Participatory Democracy during the Progressive Era*. University Park: Pennsylvania State University Press, 1998.

Miller, Zane L. "The Crisis of Civic and Political Virtue: Urban History, Urban Life and the New Understanding of the City." *Reviews in American History* 24, 3 (1996): 361.

Monkkonen, Eric H. *America Becomes Urban: The Development of U.S. Cities and Towns, 1780–1980*. Berkeley: University of California Press, 1990.

Munton, Richard. "Deliberative Democracy and Environmental Decision-Making." In *Negotiating Environmental Change: New Perspectives from Social Science*, edited by Frans Berkhout, Melissa Leach, and Ian Scoones, 109–136. Cheltenham, United Kingdom: Edward Elgar, 2003.

Nichol, Gene R. "Citizen Suits and the Future of Standing in the 21st Century: From Lujan to Laidlaw and Beyond: The Impossibility of Lujan's Project." *Duke Environmental Law and Policy Forum* 11 (Spring, 2001): 193.

Oliver, Pamela E., and Daniel J. Myers. "How Events Enter the Public Sphere: Conflict, Location, and Sponsorship in Local Newspaper Coverage of Public Events." *American Journal of Sociology* 105 (July 1999): 38–87.

Payne, Rodger A. "Freedom and the Environment." *Journal of Democracy* 6, 3 (1995): 41–55.

Pettinico, George. "Civic Participation Is Alive and Well in Today's Environmental Groups." *Public Perspective* 7, 4 (1996): 27–30.

Plattner, Marc F. "The Uses of 'Civil Society.'" *Journal of Democracy* 6, 4 (1995): 169–173.

Portz, John, Lana Stein, and Robin R. Jones. *City Schools and City Politics: Institutions and Leadership in Pittsburgh, Boston, and St. Louis*. Lawrence: University Press of Kansas, 1999.

Putnam, Robert D. *Bowling Alone: The Collapse and Revival of American Community*. New York: Simon & Schuster, 2000.

Ross, Bernard H., and Myron A. Levine. *Urban Politics: Power in Metropolitan America*. Itasca, IL: F.E. Peacock, 1996.

Ryan, Mary P. *Civic Wars: Democracy and Public Life in the American City during the Nineteenth Century*. Berkeley: University of California, 1997.

Sax, Joseph L. *Defending the Environment: A Strategy for Citizen Action*. New York: Alfred A. Knopf, 1971.

Shapiro, Michael J. "Bowling Blind: Post Liberal Civil Society and the Worlds of Neo-Tocquevillean Social Theory." *Theory & Event* 1, 1 (1997): n. p.

Skocpol, Theda. "The Tocqueville Problem: Civic Engagement in American Democracy." *Social Science History* 21 (Winter 1997): 455–479.

———. *Diminished Democracy: From Membership to Management in American Civil Life*. Norman: University of Oklahoma Press, 2003.

Skocpol, Theda, and Morris P. Fiorina, eds. *Civic Engagement in American Democracy*. Washington, DC: Brookings Institution Press, 1999.

Somers, Margaret R. "What's Political or Cultural about Political Culture and the Public Sphere? Toward an Historical Sociology of Concept Formation." *Sociological Theory* 13 (July 1995): 113–144.

Spain, Daphne. *How Women Saved the City*. Minneapolis: University of Minnesota Press, 2001.

Stone, Clarence N. *Regime Politics: Governing Atlanta, 1946–1988*. Lawrence: University Press of Kansas, 1989.

Tocqueville, Alexis de. *Democracy in America*. New York: Mentor, 1956.

Tolbert, Charles M., Thomas A. Lyson, and Michael D. Irwin. "Local Capitalism, Civic Engagement, and Socioeconomic Well-Being." *Social Forces* 77 (December 1998): 401–427.

Verba, Sidney, Kay L. Schlozman, and Henry E. Brady. *Voice and Equality: Civic Voluntarism in American Politics*. Cambridge: Harvard University Press, 1995.

Villa, Dana R. "Postmodernism and the Public Sphere." *American Political Science Review* 86 (September 1992): 712–721.

Woolcock, Michael. "Social Capital and Economic Development: Toward a Theoretical Synthesis and Policy Framework." *Theory and Society* 27 (1998): 151–208.

ENVIRONMENTAL HISTORY AND POLITICS

Alexander, Thomas G. "Cooperation, Conflict, and Compromise: Women, Men, and the Environment in Salt Lake City, 1890–1930." *BYU Studies* 35, 1 (1995): 6.

Andrews, Richard N. L. *Managing the Environment, Managing Ourselves: A History of American Environmental Policy*. 2nd ed. New Haven: Yale University Press, 2006.

Bartlett, Robert V. "Environmental History in Historical Perspective." *Journal of Policy History* 12, 3 (2000): 395–399.

Beder, Sharon. *Global Spin: The Corporate Assault on Environmentalism*. White River Junction, VT: Chelsea Green, 2002.

Beierle, Thomas C., and Jerry Cayford. *Democracy in Practice: Public Participation in Environmental Decisions*. Washington, DC: Resources for the Future, 2002.

Blum, Elizabeth D. *Love Canal Revisited: Race, Class, and Gender in Environmental Activism*. Lawrence, Kansas: University Press of Kansas, 2008.

Bosso, Christopher J. *Environment, Inc.: From Grassroots to Beltway*. Lawrence: University Press of Kansas, 2005.

Brulle, Robert J. *Agency, Democracy, and Nature: The U.S. Environmental Movement from a Critical Theory Perspective*. Cambridge, MA: MIT Press, 2000.

Buell, Frederick. *From Apocalypse to Way of Life: Environmental Crisis in the American Century*. New York: Routledge, 2003.

Carson, Rachel. *Silent Spring*. Boston: Houghton Mifflin, 1994.

Coenen, Frans H. J. M., Dave Huitema, and Laurence J. O'Toole, eds. *Participation and the Quality of Environmental Decision Making*. Dordrecht: Kluwer Academic, 2000.

Coglianese, Cary. "Litigating within Relationships: Disputes and Disturbance in the Regulatory Process." *Law and Society Review* 30, 4 (1996): 735–765.

Colton, Craig E., and Peter N. Skinner. *The Road to Love Canal: Managing Industrial Waste before* EPA. Austin: University of Texas Press, 1996.

Davis, Jack E. "'Conservation Is Now a Dead Word': Marjory Stoneman Douglas and the Transformation of American Environmentalism." *Environmental History* 8 (January 2003): 53–76.

Dowie, Mark. *Losing Ground: American Environmentalism at the Close of the Twentieth Century*. Cambridge, MA: Massachusetts Institute of Technology Press, 1995.

Dunlap, Riley E., and Angela G. Mertig, eds. *American Environmentalism: The U.S. Environmental Movement, 1970–1990*. Philadelphia: Taylor & Francis, 1992.

Egan, Michael. *Barry Commoner and the Science of Survival*. Cambridge, MA: MIT Press, 2007.

Espeland, Wendy. "Legally Mediated Identity: The National Environmental Policy Act and the Bureaucratic Construction of Interests." *Law and Society Review* 28, 5 (1994): 1149–1179.

Fischbeck, Paul S., and R. Scott Farrow, eds. *Improving Regulation: Cases in Environment, Health, and Safety*. Washington, DC: Resources for the Future, 2001.

Fischer, Frank. *Citizens, Experts, and the Environment: The Politics of Local Knowledge*. Durham, NC: Duke University Press, 2000.

Fischer, Frank, and John Forester, eds. *The Argumentative Turn in Policy Analysis and Planning*. Durham, NC: Duke University Press, 1993.

Fischoff, Baruch. "Risk Perception and Communication Unplugged: Twenty Years of Process." *Risk Analysis* 15, 2 (1995): 137–146.

Flader, Susan L. "Citizenry and the State in the Shaping of Environmental Policy." *Environmental History* 3 (January 1998): 8–24.

Flanagan, Maureen A. *Seeing with Their Hearts: Chicago Women and the Vision of the Good City, 1871–1933*. Princeton: Princeton University Press, 2002.

Freudenberg, Nicholas. *Not in Our Backyards! Community Action for Health and the Environment*. New York: Monthly Review Press, 1984.

Gilfoyle, Timothy J. "White Cities, Linguistic Turns, and Disneylands: The New Paradigms of Urban History." *Reviews in American History* 26 (March 1998): 175–204.

Gionfriddo, David P. "Sealing Pandora's Box: Judicial Doctrines Restricting Public Trust Citizen Environmental Suits." *Boston College Environmental Affairs Law Review* 13 (Spring 1986): 439.

Gosdin, John Mark. "The Environmental Impact Statement Process: The Texas Experience." Ph.D. dissertation, University of Texas, Austin, 1982.

Gottlieb, Robert. *Forcing the Spring: The Transformation of the American Environmental Movement*. Washington, DC: Island Press, 1993.

Gregory, David B. *Standing to Sue in Environmental Litigation in the United States of America*. Merges, Switzerland: International Union for Conservation of Nature and Natural Resources, 1972.

Hays, Samuel P. *Beauty, Health and Permanence: Environmental Politics in the United States, 1955–1985*. Cambridge: Cambridge University Press, 1987.

Hoberg, George. *Pluralism by Design: Environmental Policy and the American Regulatory State*. New York: Praeger, 1992.

Hoy, Suellen M. "'Municipal Housekeeping': The Role of Women in Improving Urban Sanitation Practices." In *Pollution and Reform in American Cities, 1870–1930*, edited by Martin V. Melosi, 173–198. Austin: University of Texas Press, 1980.

Hurley, Andrew. *Environmental Inequalities: Class, Race and Industrial Pollution in Gary, Indiana, 1945–1980*. Chapel Hill: University of North Carolina Press, 1995.

———, ed. *Common Fields: An Environmental History of St. Louis*. St. Louis: Missouri Historical Society Press, 1997.

June, Robert B. "Citizen Suits: The Structure of Standing Requirements for Citizen Suits and the Scope of Congressional Power." *Environmental Law* 24 (1994): 761, 779–809.

Kalen, Sam. "Standing on Its Last Legs: Bennett v. Spear and the Past and Future of Standing in Environmental Cases." *Journal of Land Use and Environmental Law* 13 (Fall 1997): 1.

Kehoe, Terence. *Cleaning Up the Great Lakes: From Cooperation to Confrontation*. Dekalb: Northern Illinois University Press, 1997.

Kline, Benjamin. *First Along the River: A Brief History of the U.S. Environmental Movement*. San Francisco: Acadia Books, 2000.

Kraft, Michael E. "U.S. Environmental Policy and Politics: From the 1960s to the 1990s." *Journal of Policy History* 12, 1 (2000): 17–42.

Landy, Marc K., Marc J. Roberts, and Stephen R. Thomas. *The Environmental Protection Agency: Asking the Wrong Questions*. Oxford: Oxford University Press, 1990.

Longhurst, James L. "'Don't Hold Your Breath, Fight for It!' Women's Activism and Citizen Standing in Pittsburgh and the United States, 1965–1975." Ph.D. diss., Carnegie Mellon University, 2004.

Mazur, Allan. *A Hazardous Inquiry: The Rashomon Effect at Love Canal*. Cambridge, MA: Harvard University Press, 1998.

McGurty, Eileen Maura. "From Nimby to Civil Rights: The Origins of the Environmental Justice Movement." *Environmental History* 2 (July 1997): 301–323.

Merchant, Carolyn. "Women of the Progressive Conservation Movement: 1900–1916." *Environmental Review* 8 (Spring 1984): 57–85.

Mercier, Jean. *Downstream and Upstream Ecologists: The People, Organizations, and Ideas behind the Movement*. Westport, CT: Praeger, 1997.

Miller, Char, and Hal Rothman, eds. *Out of the Woods: Essays in Environmental History*. Pittsburgh: University of Pittsburgh Press, 1997.

Mintz, Joel A. *Enforcement at the EPA: High Stakes and Hard Choices*. Austin: University of Texas Press, 1995.

Moore, Emmett Burris. *The Environmental Impact Statement Process and Environmental Law*. Columbus: Battelle Press, 1997.

Norwood, Vera. *Made from This Earth: American Women and Nature*. Chapel Hill: University of North Carolina Press, 1993.

Novotny, Patrick. *Where We Live, Work and Play: The Environmental Justice Movement and the Struggle for a New Environmentalism*. Westport, CT: Praeger, 2000.

Rosen, Christine Meisner, and Joel A. Tarr. "The Importance of an Urban Perspective in Environmental History." *Journal of Urban History* 20 (May 1994): 299–310.

Rothman, Hal K. *The Greening of a Nation? Environmentalism in the United States since 1945*. Fort Worth, TX: Harcourt Brace College, 1998.

Russell, Edmund P., III. "Lost among the Parts Per Billion: Ecological Protection at the United States Environmental Protection Agency, 1970–1993." *Environmental History* 2 (January 1997): 29–51.

Sellers, Christopher C. "Environmentalists by Nature: The Postwar America of Samuel Hays." *Reviews in American History* 28, 1 (2000): 112–119.

Shabecoff, Philip. *A Fierce Green Fire: The American Environmental Movement*. New York: Hill and Wang, 1993.

Simmons, W. Michelle. *Participation and Power: Civic Discourse in Environmental Policy Decisions*. Albany: SUNY Press, 2007.

Sze, Julie. *Noxious New York: The Racial Politics of Urban Health and Environmental Justice*. Cambridge, MA: MIT Press, 2007.

Taylor, Alan. "Unnatural Inequalities: Social and Environmental Histories." *Environmental History Review* 1 (1996): 6–19.

Taylor, Serge. *Making Bureaucracies Think: The Environmental Impact Statement Strategy of Administrative Reform*. Stanford, CA: Stanford University Press, 1984.

Tesh, Sylvia Noble. *Uncertain Hazards: Environmental Activists and Scientific Proof.* Ithaca: Cornell University Press, 2000.

Zinn, Matthew D. "Policing Environmental Regulatory Enforcement: Cooperation, Capture, and Citizen Suits." *Stanford Environmental Law Journal* 21 (2002): 81, 83–85.

PITTSBURGH AND PENNSYLVANIA

Alberts, Robert C. *The Shaping of the Point.* Pittsburgh: University of Pittsburgh Press, 1980.

Apelt, Brian. *The Corporation: A Centennial Biography of the United States Steel Corporation, 1901–2001.* Pittsburgh: Cathedral Publishing, University of Pittsburgh, 2000.

Bauman, John F., and Edward K. Muller. *Before Renaissance: Planning in Pittsburgh, 1889–1943.* Pittsburgh: University of Pittsburgh Press, 2006.

Bierwerth, Nancy S., Elizabeth W. Toor, and Agnes R. Tuden, eds. *Allegheny County Government: Organization, Facilities and Services.* Pittsburgh: League of Women Voters, Allegheny County Council, 1971.

Binder, Frederick Moore. *Coal Age Empire: Pennsylvania Coal and Its Utilization to 1860.* Harrisburg: Pennsylvania Historical and Museum Commission, 1974.

Boscart, Terry R., and Joel R. Burcat, eds. *Pennsylvania Environmental Law and Practice.* 3rd ed. Harrisburg: Pennsylvania Bar Institute Press, 2001.

Brignano, Mary. *The Power of Pittsburgh: History of Duquesne Light Company.* Pittsburgh: Duquesne Light Company, 1996.

Brignano, Mary, and J. Tomlinson Fort. *Reed Smith: A Law Firm Celebrates 125 Years.* Pittsburgh: Reed Smith LLP, 2002.

Crowley, Gregory J. *The Politics of Place: Contentious Urban Redevelopment in Pittsburgh.* Pittsburgh: University of Pittsburgh Press, 2005.

Eriksen, Donald P. "Air Pollution: The Problem and the Legislative and Administrative Responses of the United States, Pennsylvania and Allegheny County." *University of Pittsburgh Law Review* 30 (1968–1969): 633–670.

Ferman, Barbara. *Challenging the Growth Machine: Neighborhood Politics in Chicago and Pittsburgh.* Lawrence: University Press of Kansas, 1996.

Greenwald, Maurine W., and Margo Anderson, eds. *Pittsburgh Surveyed: Social Science and Social Reform in the Early Twentieth Century.* Pittsburgh: University of Pittsburgh Press, 1996.

Gugliotta, Angela. "Class, Gender, and Coal Smoke: Gender Ideology and Environmental Justice in Pittsburgh, 1868–1914." *Environmental History* 5 (April 2000): 165–197.

Hays, Samuel P., ed. *City at the Point: Essays on the Social History of Pittsburgh.* Pittsburgh: University of Pittsburgh Press, 1989.

Hinshaw, John. *Steel and Steelworkers: Race and Class Struggle in Twentieth-Century Pittsburgh.* Albany: State University of New York Press, 2002.

Hoerr, John P. *And the Wolf Finally Came: The Decline of the American Steel Industry.* Pittsburgh: University of Pittsburgh Press, 1988.

Krass, Peter. *Carnegie.* New York: John Wiley & Sons, 2002.

Krause, Paul. *The Battle for Homestead, 1880–1892: Politics, Culture and Steel.* Pittsburgh: University of Pittsburgh Press, 1992.

Lobes, Loretta S. "Hearts All Aflame: Women and the Development of New Forms of Social Service Organization, 1870–1930." Ph.D. dissertation, Carnegie Mellon University, 1996.

Lubove, Roy. *Twentieth-Century Pittsburgh.* 2 vols. Pittsburgh: University of Pittsburgh Press, 1995–1996.

Mershon, Sherie R. "Corporate Social Responsibility and Urban Revitalization: An Organizational History of the Allegheny Conference on Community Development, 1943–1990." Ph.D. dissertation, Carnegie Mellon University, 2000.

Misa, Thomas J. *A Nation of Steel: The Making of Modern America, 1865–1925*. Baltimore: Johns Hopkins University Press, 1995.

Snyder, Lynne Page. "'The Death-Dealing Smog over Donora, Pennsylvania': Industrial Air Pollution, Public Health, and Federal Policy, 1915–1963." Ph.D. dissertation, University of Pennsylvania, 1994.

Stave, Bruce M. *The New Deal and the Last Hurrah: Pittsburgh Machine Politics*. Pittsburgh: University of Pittsburgh Press, 1970.

Tarr, Joel A. "The Metabolism of the Industrial City—The Case of Pittsburgh." *Journal of Urban History* 28 (July 2002): 511–545.

———. *The Search for the Ultimate Sink: Urban Pollution in Historical Perspective*. Akron, OH: University of Akron Press, 1996.

———, ed. *Devastation and Renewal: An Environmental History of Pittsburgh and Its Region*. Pittsburgh: University of Pittsburgh Press, 2003.

Tarr, Joel A., and Bill C. Lamperes. "Changing Fuel Use Behavior and Energy Transitions: The Pittsburgh Smoke Control Movement, 1940–1950." *Journal of Social History* 14 (Summer 1981): 561–588.

Thomas, Clarke M. *Fortunes and Misfortunes: Pittsburgh and Allegheny County Politics, 1930–1995*. Pittsburgh: University of Pittsburgh Institute of Politics, 1997.

Toker, Franklin. *Pittsburgh: An Urban Portrait*. University Park: Pennsylvania State University Press, 1986.

Warren, Kenneth. *Big Steel: The First Century of the United States Steel Corporation, 1901–2001*. Pittsburgh: University of Pittsburgh Press, 2001.

Weber, Michael P. *Don't Call Me Boss: David L. Lawrence, Pittsburgh's Renaissance Mayor*. Pittsburgh: University of Pittsburgh Press, 1988.

Wollman, David H., and Donald H. Inman. *Portraits in Steel: An Illustrated History of Jones & Laughlin Steel Corporation*. Kent, OH: Kent State University Press, 1999.

U.S. History, Women, and the Long 1960s

Bender, Thomas. *Community and Social Change in America*. Baltimore, MD: Johns Hopkins University Press, 2000.

Berkowitz, Edward D. *Something Happened: A Political and Cultural Overview of the Seventies*. New York: Columbia University Press, 2006.

Betten, Neil, and Michael J. Austin, eds. *The Roots of Community Organizing, 1917–1939*. Philadelphia: Temple University Press, 1990.

Bloom, Alexander. *Long Time Gone: Sixties America Then and Now*. New York: Oxford University Press, 2001.

Burns, Stewart. *Social Movements of the 1960s: Searching for Democracy*. Boston: Twayne, 1990.

Caldwell, Lynton K., Lynton R. Hayes, and Isabel M. MacWhirter. *Citizens and the Environment: Case Studies in Popular Action*. Bloomington: Indiana University Press, 1976.

Cashman, Sean Dennis. *African-Americans and the Quest for Civil Rights, 1900–1990*. New York: New York University Press, 1993.

Chalmers, David. *And the Crooked Places Made Straight: The Struggle for Social Change in the 1960s*. Baltimore: Johns Hopkins University Press, 1996.

Davis, Flora. *Moving the Mountain: The Women's Movement in America since 1960*. Urbana: University of Illinois Press, 1999.

Duplessis, Rachel, and Ann Snitow. *The Feminist Memoir Project: Voices from Women's Liberation*. New York: Three Rivers Press, 1998.

Echols, Alice. *Daring to Be Bad: Radical Feminism in America, 1967–1975*. Minneapolis: University of Minnesota Press, 1990.

Epstein, Barbara. "Ecofeminism and Grass-roots Environmentalism in the United States." In *Toxic Struggles: The Theory and Practice of Environmental Justice*, edited by Richard Hofrichter, 144–151. Philadelphia: New Society, 1993.

Farber, David, and Beth Bailey, eds. *America in the Seventies*. Lawrence: University Press of Kansas, 2004.

Ferree, Myra Marx, and Patricia Yancy Martin. *Feminist Organizations: Harvest of the New Women's Movement*. Philadelphia: Temple University Press, 1995.

Fisher, Robert. *Let the People Decide: Neighborhood Organizing in America*. New York: Twayne, 1994.

Freedman, Estelle B. *No Turning Back: The History of Feminism and the Future of Women*. New York: Ballantine, 2002.

Freeman, Jo. *A Room at a Time: How Women Entered Party Politics*. Lanham, MD: Rowman & Littlefield, 2000.

———. *Social Movements of the Sixties and Seventies*. New York: Longman, 1980.

Frost, Jennifer. *An Interracial Movement of the Poor: Community Organizing and the New Left in the 1960s*. New York: New York University Press, 2002.

Gelb, Joyce, and Marian Lief Palley. *Women and the Public Interest*. Princeton: Princeton University Press, 1982.

Gibbs, Lois Marie. *Love Canal: The Story Continues*. Gabriola Island, BC: New Society, 1998.

Koven, Seth, and Sonya Michel, eds. *Mothers of a New World: Maternalist Politics and the Origins of Welfare States*. New York: Routledge, 1993.

Lawson, Steven F. *Civil Rights Crossroads: Nation, Community, and the Black Freedom Struggle*. Lexington: University Press of Kentucky, 2003.

Lear, Linda J. *Rachel Carson: Witness for Nature*. New York: Henry Holt, 1997.

Minkoff, Debra C. *Organizing for Equality: The Evolution of Women's and Racial-Ethnic Organizations in America, 1955–1985*. Philadelphia: Temple University Press, 1995.

Morris, Aldon D. *The Origins of the Civil Rights Movement: Black Communities Organizing for Change*. New York: Free Press, 1984.

Muncy, Robyn. *Creating a Female Dominion in American Reform, 1890–1935*. New York: Oxford University Press, 1991.

Payne, Charles M., and Adam Green, eds. *Time Longer than Rope: A Century of African American Activism, 1850–1950*. New York: New York University Press, 2003.

Robnett, Belinda. *How Long? How Long? African-American Women in the Struggle for Civil Rights*. New York: Oxford University Press, 1997.

Rome, Adam W. "'Give Earth a Chance': The Environmental Movement and the Sixties." *Journal of American History* 90 (September 2003): 525–554.

Skrentny, John D. *The Minority Rights Revolution*. Cambridge, MA: Belknap, 2004.

Swerdlow, Amy. *Women Strike for Peace: Traditional Motherhood and Radical Politics in the 1960s*. Chicago: University of Chicago Press, 1993.

Wilson, Bobby M. *Race and Place in Birmingham: The Civil Rights and Neighborhood Movements*. Lanham, MD: Rowman & Littlefield, 2000.

Index